Cambridge Lower Secondary

Maths

STAGE 9: STUDENT'S BOOK

Alastair Duncombe, Belle Cottingham, Rob Ellis, Amanda George,
Claire Powis, Brian Speed
Series Editor: Alastair Duncombe

Collins

William Collins' dream of knowledge for all began with the publication of his first book in 1819.

A self-educated mill worker, he not only enriched millions of lives, but also founded a flourishing publishing house. Today, staying true to this spirit, Collins books are packed with inspiration, innovation and practical expertise. They place you at the centre of a world of possibility and give you exactly what you need to explore it.

Collins. Freedom to teach.

Published by Collins
An imprint of HarperCollins*Publishers*
The News Building
1 London Bridge Street
London
SE1 9GF

HarperCollins*Publishers*
Macken House, 39/40 Mayor Street Upper,
Dublin 1, D01 C9W8, Ireland

Browse the complete Collins catalogue at
www.collins.co.uk

ISBN 978-0-00-837855-4

MIX
Paper | Supporting responsible forestry
FSC™ C007454

This book is produced from independently certified FSC™ paper to ensure responsible forest management.

For more information visit:
www.harpercollins.co.uk/green

British Library Cataloguing in Publication Data A catalogue record for this publication is available from the British Library.

Authors: Alastair Duncombe, Belle Cottingham, Rob Ellis, Amanda George, Claire Powis, Brian Speed
Series editor: Alastair Duncombe
Publisher: Elaine Higgleton
In-house project editors: Jennifer Hall and Caroline Green
Project manager: Wendy Alderton
Development editors: Anna Cox, Phil Gallagher and Jess White
Copyeditor: Alison Bewsher
Proofreader: Eric Pradel

Answer checkers: Eric Pradel and Jouve India Private Limited
Cover designer: Ken Vail Graphic Design and Gordon MacGlip
Cover illustrator: Ann Paganuzzi
Typesetter: Jouve India Private Limited
Production controller: Lyndsey Rogers
Printed and bound in India by Replika Press Pvt. Ltd.

Acknowledgements

The publishers gratefully acknowledge the permission granted to reproduce the copyright material in this book. Every effort has been made to trace copyright holders and to obtain their permission for the use of copyright material. The publishers will gladly receive any information enabling them to rectify any error or omission at the first opportunity.

p. 2 Rost9/Shutterstock; p26 Zerbor/Shutterstock; p. 90 International Olympic Committee: The Olympic Games and all Olympic Games data used in this publication are the property of the International Olympic Committee – All Rights Reserved. Used under licence;p.184 McGann Publishing/ BikeRaceInfo.com; p. 192 Vallo Zoltan/Shutterstock; p 223 and p 225 International Olympic Committee: The Olympic Games and all Olympic Games data used in this publication are the property of the International Olympic Committee – All Rights Reserved. Used under licence; p.227 Copyright © 2019 by United Nations, made available under a Creative Commons license CC BY 3.0 IGO: http://creativecommons.org/ licenses/by/3.0/igo/; p.231 From World Population Prospects 2019, Online Edition. Rev. 1., by Department of Economic and Social Affairs, Population Division (2019). Copyright © 2019, United Nations. Reprinted with the permission of the United Nations; p. 232 From World Population Prospects 2019, Online Edition. Rev. 1., by Department of Economic and Social Affairs, Population Division (2019). Copyright © 2019, United Nations. Reprinted with the permission of the United Nations; p.232 Based on free material from GAPMINDER. ORG, CC-BY LICENSE; p. 233 Yougov https://yougov.co.uk/ topics/lifestyle/articles-reports/2016/01/05/chinese-people- are-most-optimistic-world; p.234 Infographic Ratio Landmass & Population by Kate Snow Design for Bits of Science; p.242 From World Population Prospects 2019, Online Edition. Rev. 1., by Department of Economic and Social Affairs, Population Division (2019). Copyright © 2019, United Nations. Reprinted with the permission of the United Nations; p. 243 Based on free material from GAPMINDER.ORG, CC-BY LICENSE; p. 331 The Washington Examiner https://www. washingtonexaminer.com/weekly-standard/over-100-million- now-receiving-federal-welfare

Cambridge International copyright material in this publication is reproduced under licence and remains the intellectual property of Cambridge Assessment International Education.

Third-party websites and resources referred to in this publication have not been endorsed by Cambridge Assessment International Education.

With thanks to the following contributors from the first edition: Michele Conway, Sarah Sharratt, Caroline Fawcus, Deborah McCarthy and Fiona Smith.

With thanks to the following teachers and schools for reviewing materials in development: Samitava Mukherjee and Debjani Sen, Calcutta International School; Hawar International School; Adrienne Leisztinger, International School of Budapest; Sujatha Raghavan, Manthan International School; Mahesh Punjabi, Podar International School; Taman Rama Intercultural School; Utpal Sanghvi International School

Introduction

The *Collins Lower Secondary Maths Stage 9 Student's Book* covers the Cambridge Lower Secondary Mathematics curriculum framework (0862). The content has been covered in 26 chapters. The series is designed to illustrate concepts and provide practice questions at a range of difficulties to allow you to build confidence on a topic.

The authors have included plenty of worked examples in every chapter. These worked examples will lead you, step-by-step, through the new concepts. They include clear and detailed explanations. Where possible, links have been made between topics, encouraging you to build on what you know already, and to practise mathematical concepts in a different context.

Each chapter within the book contains activities and questions to help you to develop your skills with *thinking and working mathematically*. You will practise the skills of *specialising* and *generalising*, *conjecturing* and *convincing*, *characterising* and *classifying*, and *critiquing* and *improving*. You can find definitions of these characteristics on the next page. The activities and questions will help you to understand each topic. They will also develop your skills at spotting patterns and solving mathematical problems. These activities and questions are indicated by a star:

Every chapter has these helpful features:
- 'Starting point': to remind you of what you know already

- 'This will also be helpful when …': to let you know where you will use the mathematics in the future

- 'Getting started': to get you interested in the new topic through an activity

- 'Key terms' boxes: to identify new mathematical words you need to know in that chapter, and provide a definition

- Clear topic headings: so that you can see what you are going to be learning in each section of the chapter

- Worked examples: to show you how to answer questions with both formal and informal (diagrammatic) explanations provided

- 'Tip' boxes: to give you guidance on the possible methods and common errors

- Exercises: to give you practice at answering questions on each topic. The questions at the end of each exercise will be harder to stretch you

- 'Thinking and working mathematically' questions and activities (marked as   ): to help you develop your mathematical thinking. The activities will often be more open-ended in nature

- 'Think about' boxes: to suggest ideas that you might want to consider

- 'Discuss' boxes: to encourage you to talk about mathematical ideas with a partner or in class

- 'Did you know?' boxes: to explain where mathematical ideas came from and how they are applied in real life.

- 'Consolidation exercise': to give you further practice on all the topics introduced in the chapter.

- 'End of chapter reflection': to help you think about how well you have understood the ideas in the chapter, so that you can monitor your own progress.

We hope that you find this approach enjoyable and engaging as you progress through your mathematical journey.

The Thinking and Working Mathematically Star

Critiquing

Comparing and evaluating mathematical ideas, representations or solutions to identify advantages and disadvantages.
For example:

◀ Which is the best way to …?
◀ Write down the advantages and disadvantages of …

Specialising

Choosing an example and checking to see if it satisfies or does not satisfy specific mathematical criteria.
For example:

◀ Find an example of …
◀ Find … if … and …

Conjecturing

Forming mathematical questions or ideas.
For example:

◀ What would happen if …?
◀ How would you …?

Specialising and Generalising

Conjecturing and Convincing

Critiquing and Improving

Generalising

Recognising an underlying pattern by identifying many examples that satisfy the same mathematical criteria.
For example:

◀ Find a rule that connects … and …
◀ What can you conclude from …?

Convincing

Presenting evidence to justify or challenge a mathematical idea or solution.
For example:

◀ Prove that …
◀ Explain why …
◀ Show that …

Characterising and Classifying

Characterising

Identifying and describing the mathematical properties of an object.
For example:

◀ What do … have in common?
◀ Describe the properties of …

Improving

Refining mathematical ideas or representations to develop a more effective approach or solution.
For example:

◀ Find a better way to …
◀ Describe a more efficient way to …

Classifying

Organising objects into groups according to their mathematical properties.
For example:

◀ Match …
◀ Sort …
◀ Put a ring around all the … which …

Contents

1 Indices, roots and rational numbers

You will learn how to:

* Use positive, negative and zero indices, and the index laws for multiplication and division.
* Understand the difference between rational and irrational numbers.
* Use knowledge of square and cube roots to estimate surds.

Starting point

Do you remember...

* the meanings of positive and zero indices?

  For example, write the values of 3^2, 2^5 and 7^0.
* the index laws for multiplication and division?

  For example, write each expression as a single power of 9: $9^3 \times 9^5$, $9^7 \div 9^4$
* the definition of a rational number?

  For example, find all of the rational numbers in this list: $-7, -\dfrac{12}{13}, \pi, 0.\overset{\infty}{3}, 5.97$
* the values of squares and cubes of the first few integers?

  For example, write the square root of 81 and the cube root of 125.

This will also be helpful when...

* you learn the power of a power rule for indices
* you learn to simplify and manipulate surds.

1.0 Getting started

The table shows powers of 2. It includes negative indices.

$2^4 =$	$2 \times 2 \times 2 \times 2$	$= 16$	$\div$ ......
$2^3 =$			$\div$ ......
$2^2 =$			$\div$ ......
$2^1 =$			$\div$ ......
$2^0 =$			$\div$ ......
$2^{-1} =$			$\div$ ......
$2^{-2} =$			$\div$ ......
$2^{-3} =$			$\div$ ......
$2^{-4} =$			$\div$ ......

Copy the table and complete the rows for the positive and zero indices.
Now complete the labels on the arrows.

Can you use the pattern to complete the rows for the negative indices?
Describe what a negative index means.

Make a similar table for powers of 5.

Key terms

A **positive power** or **index** tells you how many copies of a number to multiply together. For example, 2^5 means $2 \times 2 \times 2 \times 2 \times 2 = 32$.

Any number to the **power of 0** equals 1.

A negative **power** or **index** tells you how many copies of a number to divide by.

For example, 2^{-3} means $1 \div 2 \div 2 \div 2$ or $\frac{1}{2 \times 2 \times 2}$, which equals $\frac{1}{8}$. This is the **reciprocal** of 2^3.

To **multiply** powers of a number, **add** the powers. For example, $3^3 \times 3^2 = 3^{3+2} = 3^5$.

To **divide** powers of a number, **subtract** the powers. For example, $3^3 \div 3^2 = 3^{3-2} = 3^1 = 3$.

A **base** is a number that has an index attached. For example, in   the base is 2.

Did you know?

There is a unit of length called a nanometre, which is 10^{-9} m or 0.000000001 m. Viruses measure about 20 to 400 nanometres across. They can only be seen with very powerful microscopes.

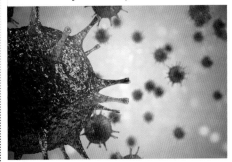

Worked example 1

Write the value of 2^{-5}

$2^{-5} = \frac{1}{2^5}$ $= \frac{1}{2 \times 2 \times 2 \times 2 \times 2}$ $= \frac{1}{32}$	2^{-5} is the reciprocal of 2^5 $2^5 = 32$, so the reciprocal of 2^5 is $\frac{1}{32}$

Worked example 2

Write each expression as a power of 4.

a) $4^2 \times 4^{-3}$ b) $4^{-2} \div 4^{-4}$ c) $(4^{-1})^3$

a) $4^2 \times 4^{-3} = 4^{2-3}$ $= 4^{-1}$	To multiply two powers of the same number, add the indices.	Longer method (without using the index rule): $4^2 = 4 \times 4$ and $4^{-3} = \dfrac{1}{4 \times 4 \times 4}$ $4 \times 4 \times \dfrac{1}{4 \times 4 \times 4} = \dfrac{4 \times 4}{4 \times 4 \times 4}$ $\qquad\qquad\qquad = \dfrac{1}{4}$
b) $4^{-2} \div 4^{-4} = 4^{-2-(-4)}$ $= 4^2$	To divide a power by another power of the same number, subtract the second index from the first index.	Longer method (without using the index rule): $4^{-2} = \dfrac{1}{4 \times 4}$ and $4^{-4} = \dfrac{1}{4 \times 4 \times 4 \times 4}$ $\dfrac{1}{4 \times 4} \div \dfrac{1}{4 \times 4 \times 4 \times 4} = \dfrac{1}{4 \times 4} \times \dfrac{4 \times 4 \times 4 \times 4}{1}$ $\qquad\qquad\qquad\qquad = \dfrac{4 \times 4 \times 4 \times 4}{4 \times 4}$ $\qquad\qquad\qquad\qquad = 4 \times 4$
c) $(4^{-1})^3 = 4^{-1} \times 4^{-1} \times 4^{-1}$ $= 4^{-1-1-1}$ $= 4^{-3}$	To cube a number, multiply three copies of the number together. Add the indices.	Longer method (without using the index rule): $4^{-1} = \dfrac{1}{4}$ $\dfrac{1}{4} \times \dfrac{1}{4} \times \dfrac{1}{4} = \dfrac{1}{4 \times 4 \times 4}$

Exercise 1

1 This table shows some powers of 2. Copy and complete the table.

Describe the pattern.

Power	2^{-4}	2^{-3}	2^{-2}	2^{-1}	2^0	2^1	2^2	2^3	2^4
Value									

2 Which of these numbers are less than 0?

5^0 $(-2)^{-2}$ 2^{-3} $(-3)^{-1}$ -4^{-2} -5^2

3 Write each number as a power of 3.

a) 9   b) $\dfrac{1}{3}$ c) 81   d) $\dfrac{1}{9}$

e) $\dfrac{1}{27}$ f) 1   g) 243   h) $\dfrac{1}{81}$

4 Rishi has written each set of numbers in order from smallest to largest, but he has made mistakes. Correct the mistakes.

    a) $2^0, 2^{-2}, 2^1, 2^{-3}$ b) $3^{-3}, 3^{-2}, 2^1, 3^0$ c) $5^{-1}, 7^{-1}, 2^{-1}, 1^{-1}$ d) $3^1, 5^{-2}, 2^2, 1^{-4}$

5 Use >, < or = to make correct statements.

    a) $3^2 \dots 2^3$ b) $10^0 \dots 3^0$ c) $5^{-2} \dots 0$ d) $2^{-5} \dots 2^{-3}$

    e) $0.5 \dots 2^{-1}$ f) $10^{-2} \dots 2^2$ g) $5^{-2} \dots 3^{-2}$ h) $2^{-4} \dots 3^{-3}$

6 Write each expression as a fraction.

    a) $2^{-2} + 2^{-1} + 2^0$ b) $3^{-1} - 3^{-2} + 3$ c) $5^0 - 5^{-2} + 5^1$ d) $4^{-2} + 2^0 - 2^{-3}$

7 Sara thinks that $9^0 > 5^0$ because the powers are the same and $9 > 5$.

Explain the mistake that Sara has made.

8 Complete the statements.

    a) $3^2 = \square^1$ b) $2^{-2} = \square^{-1}$ c) $4^{\square} = 2^6$ d) $5^{\square} = 0.04$

9 Simplify each expression. Leave your answer in index form.

    a) $3^9 \times 3^4$ b) $5^{-2} \times 5^7$ c) $4^{-4} \div 4^3$ d) $6^{-3} \times 6^{-2}$

    e) $2^6 \div 2^{-5}$ f) $7^{-5} \div 7^{-2}$ g) $5^0 \times 5^{-6}$ h) $(4^{-5})^2$

    i) $(3^{-1})^3$ j) $3^6 \times 3^{-5} \times 3^3$ k) $2^7 \times 2^{-4} \div 2^8$ l) $5^2 \div 5^4 \times 5^{-7}$

10 Simplify each expression. Leave your answer in index form.

    a) $\dfrac{3^5 \times 3^{-3}}{3^4}$ b) $\dfrac{5^{-4} \times 5^7}{5^0}$ c) $\dfrac{6^5}{6^2 \times 6^{-1}}$ d) $\dfrac{7^5 \times 7^1}{7^{-3}}$

Think about

Is this statement true or false?

$2^{-1} + 8^{-1} = 10^{-1}$

Thinking and working mathematically activity

Isabella uses two ways to show why a number to the power of zero equals 1.

$3^1 \div 3^1 = 3 \div 3 = 1$	Using an index law, $3^2 \times 3^0 = 3^{2+0} = 3^2 = 9$
Using an index law, $3^1 \div 3^1 = 3^{1-1} = 3^0$	but $3^2 = 9$
so $3^0 = 1$	so if $3^2 \times 3^0 = 9$ then $3^0 = 1$

Explore negative indices using these methods:

- Use the division $3^2 \div 3^3$ to show that $3^{-1} = \dfrac{1}{3}$

- Use the multiplication $3^{-1} \times 3^1$ to show that $3^{-1} = \dfrac{1}{3}$

- Find two ways to show that $5^{-2} = \dfrac{1}{25}$

1.2 Rational and irrational numbers

Key terms

The **natural numbers** are the positive integers: 1, 2, 3, 4, 5, ...

A **rational number** is a number that can be written as a fraction.

An **irrational number** is a number that cannot be written as a fraction.

> **Did you know?**
>
> The square root of a negative number is called an imaginary number. The square root of negative one, $\sqrt{-1}$, has the symbol i in mathematics. Imaginary numbers are different from real numbers. They are very useful for doing calculations in some areas of science, such as electrical engineering and quantum physics.

Worked example 3

Here is a list of numbers.

$$\sqrt{18} \qquad 0.1\dot{6} \qquad 0.867 \qquad -5 \qquad \sqrt[3]{6} \qquad 1\frac{6}{7} \qquad \pi \qquad \sqrt{-4} \qquad \sqrt[3]{8} \qquad \sqrt[3]{-8}$$

Write down the numbers that are:

a) rational

b) irrational

c) neither rational nor irrational.

a)	$0.1\dot{6}$	All recurring decimals can be written as fractions: $0.1\dot{6} = \frac{1}{6}$
	0.867	Any terminating decimal can be written as a fraction: $0.867 = \frac{867}{1000}$
	-5	Any integer can be written as a fraction: $-5 = -\frac{5}{1}$
	$1\frac{6}{7}$	Any mixed number can be written as an improper fraction: $1\frac{6}{7} = \frac{13}{7}$
	$\sqrt[3]{8}$	$\sqrt[3]{8} = 2$, which is rational.
	$\sqrt[3]{-8}$	$\sqrt[3]{-8} = -2$, which is rational.
b)	$\sqrt{18}$	The square root of any number that is not a square number is irrational.
	$\sqrt[3]{6}$	The cube root of any number that is not a cube number is irrational.
	π	π cannot be written exactly as a fraction. It is a decimal that neither terminates nor recurs.
c)	$\sqrt{-4}$	The square root of a negative number is neither rational nor irrational.
		You will not be expected to work with numbers like these.

Thinking and working mathematically activity

Which Venn diagram correctly shows the relationships between natural numbers (N), integers (Int), rational numbers (R) and irrational numbers (Irr)? Explain your answer.

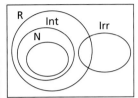 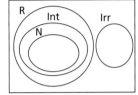 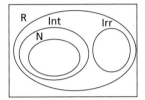

Make a larger copy of the correct Venn diagram. Write the following numbers in the correct regions of your diagram.

$\sqrt{5}$ $\dfrac{8}{9}$ $\sqrt{16}$ -6 22 $\sqrt[3]{27}$ $5.\overset{\infty}{2}$ 19 π $-\sqrt{9}$ $\sqrt[3]{-27}$

Exercise 2

1–5, 7

1 Here is a list of numbers.

-7.2 $\sqrt{8}$ -21 $\dfrac{13}{11}$ $-\sqrt{4}$

Write down a number that is:

a) both rational and an integer

b) rational but not an integer

c) irrational.

2 Here is a list of numbers.

-5 $\dfrac{22}{7}$ $\sqrt{12}$ 14.75 $\sqrt{81}$ $0.\overset{\infty}{3}$ π

Sort the numbers into rational and irrational numbers.

3 Sort these numbers into rational and irrational numbers.

$\sqrt{4}$ $\sqrt{5}$ $\sqrt{6}$ $\sqrt{7}$ $\sqrt{8}$ $\sqrt{9}$

4 Give an example where:

a) the product of two irrational numbers is a rational number

b) the product of two rational numbers is a rational number

c) the product of a rational number and an irrational number is an irrational number.

5 Decide whether each statement is true or false. If it is true, give an example.

a) An integer can be an irrational number.

b) An irrational number can be negative.

c) The square root of a natural number can be a natural number.

d) The square root of a natural number can be a negative integer.

e) The square root of a natural number can be a rational number.

f) The square root of a natural number can be an irrational number.

6 Use a calculator to find which of the values are rational and which are irrational.

a) $\sqrt{500}$ b) $\sqrt{289}$ c) $\sqrt{27}$ d) $\sqrt{3} \times \sqrt{12}$

e) $\sqrt[3]{-729}$ f) $\sqrt{361}$ g) $\sqrt[3]{25}$ h) $\sqrt[3]{1331}$

7 Vocabulary question Copy and complete the sentences below using terms from the box.

recurring   squares   rational   terminating   natural   square roots   irrational

Numbers that can be written as fractions are called .... numbers. These include ....

decimals and .... decimals. Numbers that cannot be written as fractions are called

.... numbers. These include .... of numbers that are not perfect ....

1.3 Estimating surds

Key terms

A **surd** is a square root of a number that is not a square number, or a cube root of a number that is not a cube number. For example $\sqrt{2}$, $\sqrt{12}$ and $\sqrt[3]{15}$ are surds, but $\sqrt{4}$ and $\sqrt[3]{27}$ are not surds. Surds are irrational.

Worked example 4

Estimate:

a) $\sqrt{55}$

b) $\sqrt[3]{20}$

a) $7^2 = 7 \times 7 = 49$ $8^2 = 8 \times 8 = 64$ $49 < 55 < 64$ $\sqrt{49} < \sqrt{55} < \sqrt{64}$ $7 < \sqrt{55} < 8$ $\sqrt{55}$ is about 7.4 (or 7.5)	Use square number facts: 55 is between $7^2 = 49$ and $8^2 = 64$. 55 is closer to 49 than to 64, so $\sqrt{55}$ is closer to 7 than to 8.	
b) $2^3 = 8$ $3^3 = 27$ $8 < 20 < 27$ $\sqrt[3]{8} < \sqrt[3]{20} < \sqrt[3]{27}$ $2 < \sqrt[3]{20} < 3$ $\sqrt[3]{20}$ is about 2.7 (or 2.8)	Use cube number facts: 20 is between $2^3 = 8$ and $3^3 = 27$. 20 is closer to 27 than to 8, so $\sqrt[3]{20}$ is closer to 3 than to 2.	

1. Write down the value of:

   a) $\sqrt{81}$　　　　b) $\sqrt{144}$　　　　c) $\sqrt{256}$

   d) $\sqrt[3]{1}$　　　　e) $\sqrt[3]{64}$　　　　f) $\sqrt[3]{125}$

2. a) Copy and complete this number line.

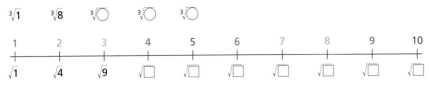

$\sqrt[3]{1}$　　$\sqrt[3]{8}$　　$\sqrt[3]{\bigcirc}$　　$\sqrt[3]{\bigcirc}$　　$\sqrt[3]{\bigcirc}$

| 1 | 2 | 3 | 4 | 5 | 6 | 7 | 8 | 9 | 10 |

$\sqrt{1}$　$\sqrt{4}$　$\sqrt{9}$　$\sqrt{\square}$　$\sqrt{\square}$　$\sqrt{\square}$　$\sqrt{\square}$　$\sqrt{\square}$　$\sqrt{\square}$　$\sqrt{\square}$

   b) Use the number line to estimate a value for:

   i) $\sqrt{40}$　　　　**ii)** $\sqrt{70}$　　　　**iii)** $\sqrt{93}$

   iv) $\sqrt[3]{35}$　　　**v)** $\sqrt[3]{54}$　　　**vi)** $\sqrt[3]{100}$

3. Is each statement true or false?

   a) $6 < \sqrt{44} < 7$　　b) $9 < \sqrt{104} < 10$　　c) $11 < \sqrt{120} < 12$　　d) $11 < \sqrt{125} < 12$

   e) $2 < \sqrt[3]{3} < 3$　　f) $3 < \sqrt[3]{11} < 4$　　g) $5 < \sqrt[3]{29} < 6$　　h) $3 < \sqrt[3]{45} < 4$

4. Estimate each value to one decimal place.

   a) $\sqrt{30}$　　　b) $\sqrt{68}$　　　c) $\sqrt{117}$　　　d) $\sqrt{122}$

   e) $\sqrt[3]{10}$　　　f) $\sqrt[3]{24}$　　　g) $\sqrt[3]{70}$　　　h) $\sqrt[3]{113}$

5. $\square$ is an integer.

   Find all of the possible values of $\square$ if:

   a) $1 < \sqrt{\square} < 2$　　　b) $2 < \sqrt{\square} < 3$　　　c) $1 < \sqrt[3]{\square} < 2$

6 Points A, B, C, D and E on the number line below correspond to one number from the list below:

$\sqrt{2}$　　$\sqrt[3]{1}$　　$\sqrt{5}$　　$\sqrt[3]{30}$　　$\sqrt{23}$

Match each number above with one of the points on the number line.

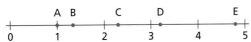

```
    A  B      C      D          E
 ---+--+---+--+---+--+---+--+---+--+---
    0  1      2      3      4      5
```

7. Is each statement true or false?

   a) $\sqrt{19} < \sqrt[3]{60}$　　　b) $\sqrt[3]{100} > \sqrt{8}$　　　c) $8.2^2 < 64$　　　d) $\sqrt{80} < 2.7^3$

Thinking and working mathematically activity

How would you use a calculator to find $\sqrt{200}$ without using the square root button? What inverse operation would you use?

- Find $\sqrt{200}$ correct to 2 decimal places. Show every step in your working.

- How do you know your answer is correct to 2 decimal places?

- Without using the cube root button, use a calculator to find $\sqrt[3]{200}$ correct to 2 decimal places. Show your working.

- Write a method for finding a root without using a root button.

Consolidation exercise

1 Is each statement true or false? Correct the statements that are incorrect.

a) $5^{-2} = -25$ b) $3^{-3} = \dfrac{1}{9}$ c) $1^{-2} = \dfrac{1}{2}$ d) $10^{-1} = 0.1$

2 Use the numbers -4, -3, -2 and -1 once each to complete these statements.

a) $6^{\square} = \dfrac{1}{36}$ b) $4^{\square} = 0.25$ c) $1^{\square} = 1^4$ d) $10^{\square} = 0.0001$

3 For the calculation $4^{-2} \times 4^{-2} \times 4^{-2}$, Atuno wrote $4^{(-2) \times (-2) \times (-2)} = 4^{-8}$. Explain why he is incorrect. Find the correct answer.

4 Find the missing numbers.

a) $4^{\square} \times 4^6 = 4^{-2}$ b) $(7^5)^2 = 7^{\square}$ c) $4^3 \times 2^2 = 4^{\square}$ d) $3^{-8} \div 9^{\square} = 3^{-12}$

> Tip
>
> In question 4 part c, rewrite 2^2 as a power of 4.
> In part d, what power of 3 would you divide 3^{-8} by to get 3^{-12} ?
> Can you rewrite it as a power of 9?

5 Use the clues to find the value of the number n.

a) n is an irrational number. It is the square root of a natural number that is greater than 7 and less than 10.

b) n is a natural number greater than $\sqrt{15}$ and less than 5.

6 a) Write down a rational number between 3 and 4.

b) Write down an irrational number between 3 and 4.

c) Write down a rational number between $\sqrt{5}$ and $\sqrt{10}$.

d) Write down an irrational number between $\sqrt{5}$ and $\sqrt{10}$.

7 Is each statement true or false?

   a) $\sqrt{10} > 3$ b) $11^3 < 1000$ c) $\sqrt{20} < 4$ d) $2.2^3 > 8$

8 Copy the number line. Estimate the position of these numbers.

   a) $\sqrt{44}$ b) $\sqrt{94}$ c) $\sqrt[3]{31}$ d) $\sqrt[3]{110}$

```
0   1   2   3   4   5   6   7   8   9   10  11  12  13  14  15
+---+---+---+---+---+---+---+---+---+---+---+---+---+---+---+
```

9 Write down a number with one decimal place that is a good approximation of:

   a) $\sqrt{200}$ b) $\sqrt[3]{200}$

10 The area of a square is 115 m². Estimate its perimeter.

11 Use two consecutive integers to make these statements correct.

   a) .... $< \sqrt{85} <$ ....   b) .... $< \sqrt[3]{20} <$ ....

> **Tip**
>
> Consecutive integers are numbers that follow on from one another in order, such as 2 and 3, 34 and 35, 609 and 610, etc.

End of chapter reflection

You should know that...	You should be able to...	Such as...
Indices can be negative.	Find the value of a number with a negative index. Use the index laws for multiplication and division with positive, zero and negative indices.	Write the value of 3^{-2} Write as a single power: a) $6^{-3} \times 6^7$ b) $11^{-5} \div 11^{-4}$
Irrational numbers are numbers that cannot be written as fractions. The square root of a number that is not a perfect square is irrational. It is called a surd.	Recognise irrational numbers, including surds.	Sort the numbers into rational and irrational. -5.28, $\sqrt{8}$, $\frac{15}{38}$, $\sqrt[3]{-64}$, π, $\sqrt[3]{18}$, $0.2\overset{\infty}{3}$
You can estimate the value of a surd by finding the two square (or cube) numbers closest to the number under the square (or cube) root.	Estimate the value of a surd, stating which whole number it is closest to.	a) Estimate the value of $\sqrt{45}$ b) Estimate the value of $\sqrt{69}$

2 Angles

You will learn how to:

- Derive and use the formula for the sum of the interior angles of any polygon.
- Know that the sum of the exterior angles of any polygon is 360°.
- Use properties of angles, parallel and intersecting lines, triangles and quadrilaterals to calculate missing angles.

Starting point

Do you remember…

- that corresponding, alternative and vertically opposite angles are equal?

  For example, find the sizes of angles marked *x* and *y*.

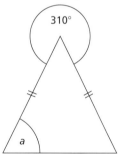

- that angles in a triangle add up to 180° and that angles in a quadrilateral add up to 360°?
  For example, find the sizes of the angles marked *a* and *b*.

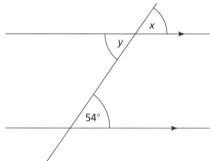 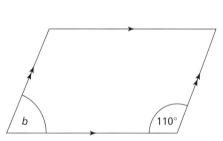

- how to use properties of the exterior angle of a triangle?
  For example, calculate the size of angle *u*.

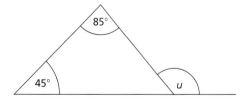

This will also be helpful when…

- you learn to solve problems that involve bearings.

Look carefully at the diagrams below. They show the exterior angles of a triangle and of a hexagon.

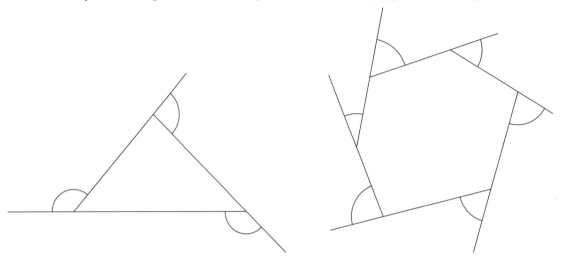

- Draw several triangles and find the sum of the three exterior angles for each triangle.
- Make a conclusion about the sum of the exterior angles in a triangle. Try to explain why your result is true.
- Investigate the sum of the exterior angles in other polygons.

Think about

Imagine starting at the marked point on the hexagon.

Imagine a journey around the outside of the hexagon.

Which angles do you turn through on the journey?

What is the total angle you will have turned through?

START

2.1 Interior and exterior angles

Key terms

The **interior angles** of a polygon are the angles at the vertices inside the polygon (shaded yellow on the diagram).

The **exterior angles** of a polygon are the angles outside the polygon formed by extending the edges of the polygon at the vertex (shaded green on the diagram).

The sum of an interior angle and an exterior angle at any vertex is 180°.

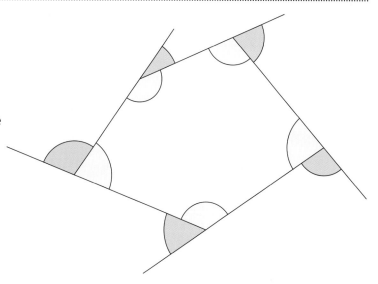

Thinking and working mathematically activity

Look at the polygon below: see how many sides each has and how many triangles you can split the shape into.

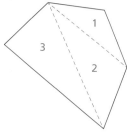

- The sum of the interior angles is 540°. Use the diagram above to help you explain why.

- Draw other polygons. Split each polygon into triangles. Investigate how the number of triangles is related to the number of sides in the polygon.

- Find a rule for finding the sum of the interior angles of a polygon with different numbers of sides.

Worked example 1

a) Find the size of the angle marked *t*.

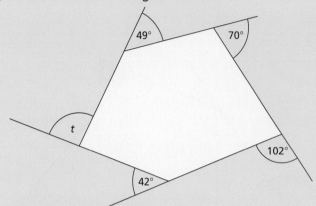

b) Find the sum of the interior angles in a hexagon.

c) Find the value of *x* in the shape below.

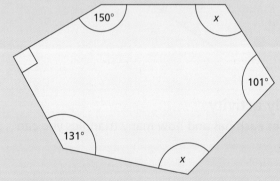

a) The sum of the four given angles is: 49° + 70° + 102° + 42° = 263° So, *t* = 360° − 263° = 97°	The sum of the exterior angles of **any** polygon is 360°. Subtract the four angles that are known from 360° to find the missing one.	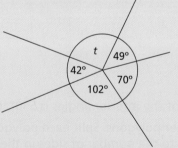
b) The sum of the interior angles in a hexagon is (6 − 2) × 180° = 4 × 180° = 720°	Remember that a polygon with *n* sides can be divided into (*n* − 2) triangles. So, the sum of interior angles in a shape with *n* sides is (*n* − 2) × 180°	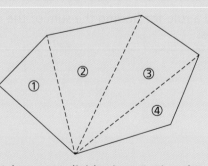 A hexagon divides into 4 triangles.

c) The sum of the four given angles is:

$90 + 150 + 101 + 131$
$= 472°$

The sum of all the interior angles in a hexagon is 720°.

$2x = 720 - 472 = 248°$

So, $x = 124°$

Begin by finding the sum of the given angles.

Write down the sum of all the interior angles in a hexagon.

Use this sum to find the total of the two unknown angles.

Halve this to find the value of x.

720°					
90°	150°	101°	131°	x	x
472°				$2x = 248°$	
				124°	124°

Exercise 1

The diagrams in this exercise are not to scale.

1 Find the size of the lettered angles in these diagrams.

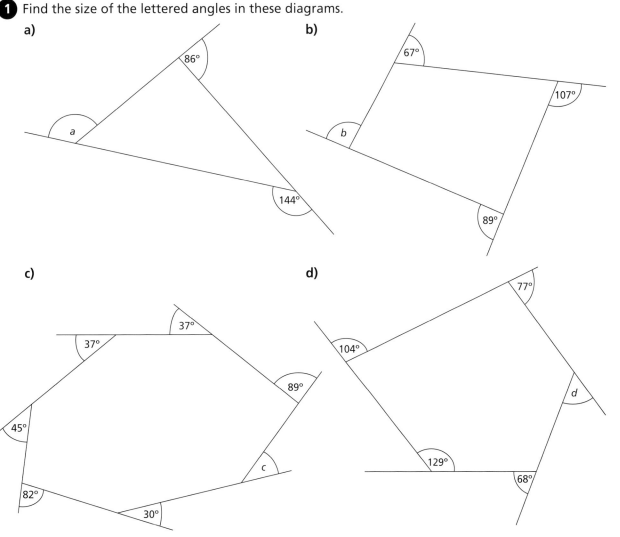

a)

86°

a

144°

b)

67°

107°

b

89°

c)

37°

37°

89°

45°

c

82°

30°

d)

77°

104°

d

129°

68°

2 Find the size of angle *x*.

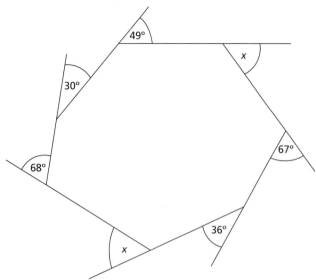

3 Calculate the sum of the interior angles of:

a) a nonagon   **b)** a 12-sided polygon   **c)** a 16-sided polygon.

4 Calculate the size of the missing angles in these polygons.

a)

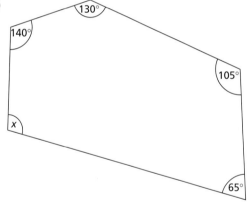

b)

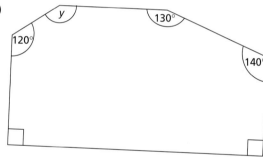

c)

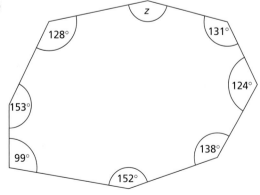

d)

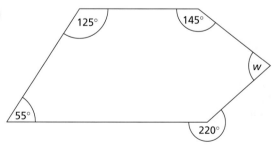

5 Calculate the value of x.

a)

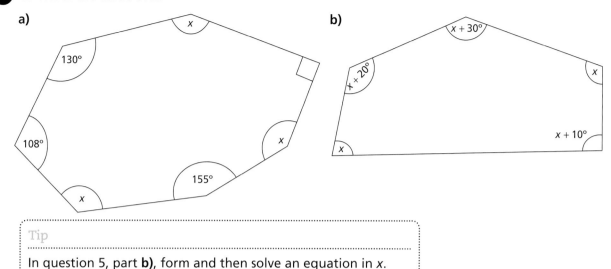

b)

Tip

In question 5, part **b)**, form and then solve an equation in x.

6 How many sides does a polygon have if its interior angles add up to 3240°?

7 Find which of these could not be the sum of the interior angles of a polygon. Give a reason for your answer.

   1260°   1800°   2520°   5640°

8 Sophia draws a polygon. None of the angles in Sophia's polygon are reflex.

She measures all the angles of the polygon, except one.

The sum of the angles she measures is 987°.
Find the size of the missing angle. Show how you worked out your answer.

9 Three of the angles in a hexagon are each 150°. The remaining three angles are in the ratio 2 : 3 : 4.

Find the size of the smallest angle.

2.2 Angles of a regular polygon

Key terms

A **regular polygon** has all its interior angles equal and all its exterior angles equal.

A shape can be **tessellated** if it can be used to make a repeating pattern with no overlaps or gaps. A tessellation made from regular polygons is called a **regular tessellation**.

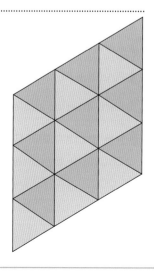

Worked example 2

a) i) What is the size of the exterior angle of a regular hexagon?

   ii) What is the size of the interior angle of a regular hexagon?

b) The size of each exterior angle in a regular polygon is 22.5°.
Find the number of sides the polygon has.

a) i) The sum of the exterior angles of a hexagon is 360°. Each exterior angle is 360° ÷ 6 = 60°	The sum of the exterior angles of any polygon is 360°. A regular hexagon has six equal exterior angles. To find the size of each one, you divide 360° by 6	360° 60° 60° 60° 60° 60° 60°
b) ii) The interior angle is 180° − 60° = 120°	The interior and exterior angles form a straight line. They add up to 180°.	60° 120° 60° 120°
c) The number of sides is 360° ÷ 22.5° = 16	The exterior angles of any polygon add up to 360°. Find how many times 22.5° divides into 360°	360° 22.5° 22.5° ...... 22.5° 22.5° 16 angles

Exercise 2

1 a) Find the size of the exterior angle of a regular nonagon.

   b) Find the size of the interior angle of a regular nonagon.

2 a) Find the size of each exterior angle of a regular polygon with 30 sides.

   b) Find the size of each interior angle of a regular polygon with 30 sides.

3 A dodecagon has 12 sides. Find the size of each interior angle of a regular dodecagon.

4 Find the numbers of sides of a regular polygon if each exterior angle is: a) 24°   b) 10°

5 Each interior angle of a regular polygon is 165°. Find the number of sides.

6 Leanne said that she measured the interior angle of a regular polygon as 160°.

Could she be correct? Explain your answer.

7 Find the name of the regular polygon with each of these properties.

    a) the interior angle is equal to the exterior angle

    b) the exterior angle is double the interior angle

    c) the sum of the interior angles is three times the sum of the exterior angles

    d) the exterior angle is not an integer, but just bigger than 50°.

8 The diagram shows a regular pentagon, a regular octagon and a rhombus.

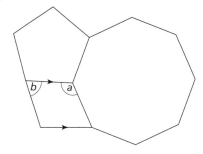

    a) Show that $a = 117°$.

    b) Write down the size of angle b.

Thinking and working mathematically activity

- Find three regular polygons that will tessellate.
- Explain why a regular decagon will not tessellate.
- Explain why the three regular polygons that you found that tessellate are the only regular polygons that tessellate.

2.3 Solving geometrical problems

Worked example 3

ABE is an isosceles triangle with $AB = AE$.

BDE is an isosceles triangle with $BE = DE$.

ABC and ED are parallel lines.

Angle $BED = 44°$

Find   **a)** angle EAB **b)** angle ABD.

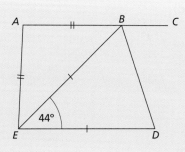

a) Ange $ABE = 44°$ Angle $AEB = 44°$ So, angle $BAE = 180 - 44 × 2$ $= 92°$	Angle ABE = Angle BED (alternate angles). Angle ABE = angle AEB (base angles of an isosceles triangle). The angles in a triangle add up to 180°	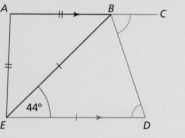
b) Angle $BDE = \dfrac{180 - 44}{2}$ $= 68°$ Angle $CBD = 68°$ Angle $ABD = 180 - 68$ $= 112°$	Angle EBD = angle BDE (base angles of an isosceles triangle are equal). Angles in a triangle add up to 180°. Angle CBD = angle BDE (alternate angles). Angles on a straight line add up to 180°	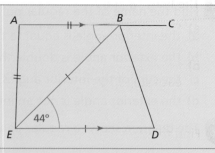

Exercise 3

1 Find the size of the angles marked with letters in each diagram.

a)
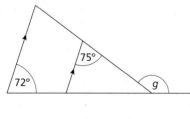

b)
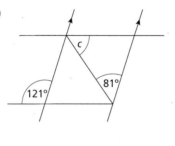

c)
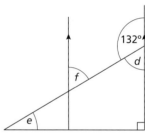

d)
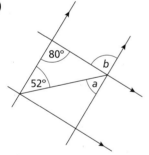

e)
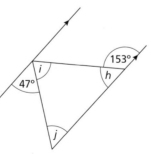

f)
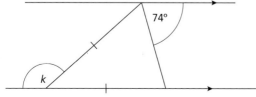

2 The lines *EB* and *DC* are parallel.

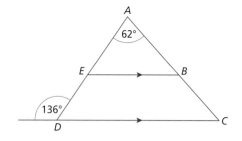

a) Find angle *AEB*. Give geometrical reasons for each step of your working out.

b) Find angle *ACD*. Give geometrical reasons for each step of your working out.

3 Find the sizes of angles *x* and *y*.

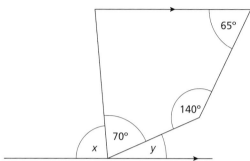

4 In the diagram, *ABCD* is a trapezium with *BC* parallel to *AD*. *E* is a point on *AD*.

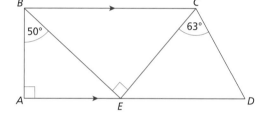

Angles *BAE* and *BEC* are right angles.

Angle *ABE* = 50° and angle *ECD* = 63°.

Find the sizes of the following angles.
Give geometrical reasons for your answers.

a) angle *BCE* b) angle *CDE*

5 *ABCD* is a rectangle. Find the values of *a* and *b*. Give geometrical reasons for each step of your answers.

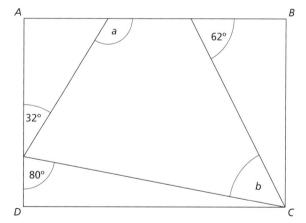

6 In the diagram, *ABC* is a right-angled triangle.

Arianne says, 'Angle *BCD* must be 25° because it is the same as the angle next to it.' Arianne is not correct.

Explain how she could find angle *BCD*.

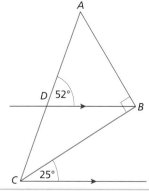

7 *BCDE* is a kite. Find the value of *t*.

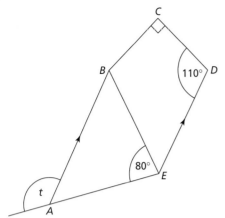

Thinking and working mathematically activity

Write an angle question for your partner to solve. Draw a diagram with a missing angle, *x*. The answer should be 48°.

You should make a problem that:

- has 2 or 3-steps
- involves several different angle facts or rules (such as angles in parallel lines, triangles or quadrilaterals).

Write a solution to your problem. Remember to give a geometric reason for each step in your solution.

Look to see if there is more than one way to solve your problem. Write a solution for each method.

Consolidation exercise

1 **a)** Write down the sum of the exterior angles of a pentagon.

b) Find the size of the angles marked by letters in these diagrams.

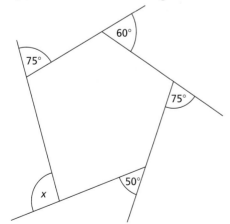

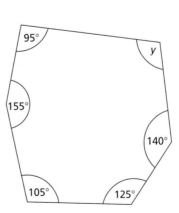

2 **a)** Jana says that the sum of the exterior angles of an octagon must be bigger than the sum of the exterior angles of a hexagon because an octagon has two more exterior angles that a hexagon. Is she correct? Explain your answer.

b) Jana says that the sum of the interior angles of an octagon must be 360° bigger than the sum of the interior angles of a hexagon because an octagon can be divided into two more triangles than a hexagon. Is she correct? Explain your answer.

3 Vocabulary question Copy and complete the sentences using some of the words in the box.

square	60°	pentagon	decagon	triangles	polygon
regular	octagon	180/n	360/n	120°	

a) A regular figure with all its exterior angles measuring 90° is a ....

b) Each interior angle of a regular hexagon is ....

c) To find the sum of the interior angles of a .... you divide it into ....

d) Each interior angle of a regular .... is 144°.

e) Each exterior angle of a regular *n*-sided figure is .... degrees.

4 Here is the diagram of a roof and its support struts.

Find the values of angles *x*, *y* and *z*. You should explain each step of your reasoning using a geometrical rule or fact.

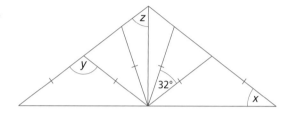

5 Mandeep said he had measured an external angle of a regular polygon as 35°.

His measurement is incorrect.

Explain how you can tell he is incorrect.

6 Calculate how many sides a regular polygon has if each interior angle is 168°.

7 Explain why a regular nonagon will not tessellate.

8 Calculate the size of the lettered angles.

a)

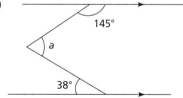

b)

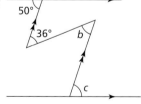

End of chapter reflection

You should know that...	You should be able to...	Such as...
The interior angles of a polygon with *n* sides add up to $(n-2) \times 180°$ The exterior angles of a polygon add up to 360°.	Find the sum of the interior angles of any polygon. Find missing interior or exterior angles of a polygon when all the others are known.	Find the sum of the interior angles of an octagon. What is the angle marked *x* in this diagram?
In a regular polygon, all interior angles are equal and all exterior angles are equal. Equilateral triangles, squares and regular hexagons are the only regular polygons that tessellate.	Find the interior angle of a regular polygon. Find the exterior angle of a regular polygon.	Find the exterior angle of a regular 10-sided shape. Find the interior angle of a regular polygon with 40 sides.
Geometric reasons may need to be given when finding the size of missing angles. These could include: • alternate angles and corresponding angles are equal • vertically opposite angles are equal • angles on a straight line add up to 180° • angles in a triangle add up to 180° • angles in a quadrilateral add up to 360°	Use properties of angles in parallel and intersecting lines, triangles and quadrilaterals to calculate missing angles.	*ABC* is an equilateral triangle. Calculate the size of the angle marked *x*.

3 Collecting and organising data

You will learn how to:

- Select, trial and justify data collection and sampling methods to investigate predictions for a set of related statistical questions, considering what data to collect, and the appropriateness of each type (qualitative or quantitative; categorical, discrete or continuous).
- Explain potential issues and sources of bias with data collection and sampling methods, identifying further questions to ask.
- Record, organise and represent categorical, discrete and continuous data. Choose and explain which representation to use in a given situation:
  o Venn and Carroll diagrams
  o tally charts, frequency tables and two-way tables

• •

Starting point

Do you remember…

- how to select and justify a suitable data collection method?

  For example, the manager of a hotel wants to find out what her customers think about the quality of the hotel's bedrooms. Select a suitable way for the manager to collect the information. Give a reason for your choice.

- how to select and justify a suitable data sampling method?

  For example, a bus company owns 300 buses. The company wants to check the tyres on a sample of 30 buses. How could the bus company choose a suitable sample? Give a reason for your choice.

- that inequality symbols can be used to define intervals for continuous data?

  For example, copy and complete the first column of the frequency table to give four equal width intervals.

Length, L (cm)	Frequency
$20 < L \leq 25$	
....	
....	
....	

This will also be helpful when…

- you learn about drawing frequency polygons to illustrate data in a frequency table
- you find an estimate of the mean of a grouped frequency table.

3.0 Getting started

Erica is researching the effects of climate change.

She plans to measure the length and width of a sample of leaves of a particular tree.

She wants to investigate how the measurements change over the years.

Discuss how Erica could obtain a sample of leaves from the tree.

What challenges could she face in getting a sample?

What are the advantages and disadvantages of different methods?

3.1 Data collection

Key terms

Quantitative data consists of numerical values. Examples include:

- age
- height
- time

Quantitative data can be discrete or continuous.

Qualitative data consists of non-numerical values and is typically descriptive. Examples include:

- hair colour
- birth place
- written feedback on homework ('excellent work', 'remember to show your working', ...).

Some data collection methods can give information that is **biased**. **Bias** can result if some groups of people are over-represented or under-represented in a survey.

Worked example 1

Mateo wants to compare the amount of time that males and females spend watching television each day.

a) What type of data is 'time watching television'?

b) Mateo asks 22 boys and 8 girls from his class how much television, to the nearest minute, they watch on an average day. Comment on Mateo's data collection method.

---

a) Mateo's data is **quantitative** as 'time watching television' is a numerical value. Time is also a **continuous** quantity.

---

b) Mateo's data is likely to be **biased** because he is collecting data only from students in his class. He should be collecting data from people of all ages.

Mateo is not asking enough people and is asking many more males than females.

People will not remember how much television they watched to the nearest minute.

Mateo should collect data from a sample of people who are representative of all people he is interested in seeking information about. People of different ages may watch different amounts of television.

Mateo's sample size should be large enough to ensure his results are accurate. As Mateo wants to compare males with females, he should ask similar numbers of each gender.

The accuracy of measurement should be realistic for the context. Asking for the time to the nearest 15 minutes or 30 minutes would be more appropriate.

Exercise 1

1 Write whether each of these variables is quantitative or qualitative.

   a) Item of clothing (jumper, shirt, sock, …)

   b) Temperature (measured in °C)

   c) Number of visitors to an exhibition

   d) Title of book

   e) Cost of items in a supermarket.

2 Mona is a clothes designer. She wants to collect information about the height of people to help her design a dress. She could collect height data using one of these methods.

Method A	**Method B**
Measure the height of a sample of people using a tape measure.	Look at a sample of people and decide if each person is very short, short, medium height, tall or very tall.

   a) Write down for each method whether Mona would be collecting qualitative or quantitative data.

   b) Give a reason why Mona may want to use Method A.

   c) Give a reason why Mona may want to use Method B.

   d) Which method would you recommend for Mona to use? Give a reason for your choice.

3 Jacqui works in a garage. She is trying to design a form that can be filled in when a customer brings their car into the garage for repairs.

She wants to collect information about the condition of each car.

Here are two variables she is considering recording.

Variable A	**Variable B**
Number of scratches on the car	Description of car's condition

a) What type of data is Variable A?

b) What type of data is Variable B?

c) Which of the variables do you think Jacqui would find more appropriate? Give a reason for your choice.

4 Florence wants to compare the number of text messages sent one week by men and women.

a) Write down the two variables that Florence will need to record for her investigation. For each one, say whether the data are qualitative or quantitative.

b) She sends a text message to the people in her address book to ask her friends how many text messages they sent last week. Why might this not give her reliable results?

5 The manager at a swimming pool wants to know what people think of his pool. He needs the information quickly so he asks 20 people using his pool one Wednesday morning.

a) Give a reason why his data collection method might not give him accurate data.

b) Suggest a better way for the manager to collect the information he needs.

> **Think about**
>
> In question 5, the manager is collecting primary data. Why is a primary source of data more suitable in this situation than a secondary source?

6 Greg wants to find out the opinions of factory workers about the factory gym.

He interviews 50 people using the gym on different days during a one week period.

Give a reason why the data Greg obtains may not accurately represent the opinions of all factory workers.

> **Discuss**
>
> How could Greg get more accurate information?

7 Yuri wants to find out how students in his school feel about homework.

He designs a questionnaire and gives it to some of the people in his year.

His questionnaire contains the following question:

Don't you agree that we get too much homework?

Write down the possible sources of bias in Yuri's investigation.

8 A shop has 80 full-time and 240 part-time employees.

The shop wants to ask its employees what they think about their pay.

The manager asks 20 full-time and 12 part-time employees.

a) Explain why the manager has not chosen a fair sample.

b) Suggest a better way for the manager to get the sample.

▼ Thinking and working mathematically activity

A pizza restaurant wants to create an advertisement that will persuade people to go there.

It hopes to collect some data that it can use to show how much customers love their pizzas.

Pia's Pizzas

More than 9 out of 10 of our customers love our pizzas

- Design a way that the restaurant could collect deliberately biased data from customers that would appear to show that people think positively about their pizzas.

- Explain why your method will give misleading information.

- Explain how unbiased information could be collected.

Think about

Andrei sees this advert:

Discuss how Andrei could collect data to see whether the claim in the advert is correct.

Supa Batteries

Tests prove that our batteries last 50% longer than rival makes.

Buy the best.

3.2 Frequency tables and Venn diagrams

Key terms

A **data collection sheet** can be used to record data from a survey or experiment. For example, this data collection sheet could be used to record the times that boys and girls took to complete a task.

Time, t (min)	$1 \leq t < 1.5$	$1.5 \leq t < 2$	$2 \leq t < 2.5$	$2.5 \leq t < 3$	$3 \leq t < 3.5$
Boys					
Girls					

Worked example 2

Martha wants to collect data on the wingspans of greenfinches (a type of bird). She expects the birds to have a wingspan of between 24 cm and 28 cm. Design a data collection sheet with four class intervals that would be suitable for collecting Martha's data. All class intervals should have equal width.

Wingspan, x (cm)	Tally	Total
$24 \leq x < 25$		
$25 \leq x < 26$		
$26 \leq x < 27$		
$27 \leq x < 28$		

The data collection sheet needs four rows, so split the interval between 24 cm and 28 cm into four classes:

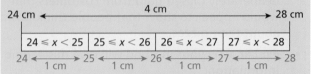

The data being collected are continuous so each class interval should be written with inequality symbols.

Worked example 3

The table shows information about the number of cars for sale in a garage.

	Cars less than 3 years old		Cars 3 or more years old	
	with air conditioning	without air conditioning	with air conditioning	without air conditioning
Petrol engine	11	5	4	6
Diesel engine	7	2	3	9

Show this information as a Venn diagram.

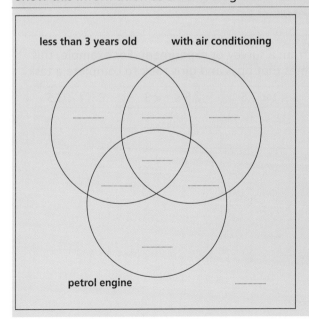

Begin by drawing the circles for the Venn diagram.

Three circles are needed as there are three categories:

- type of engine (petrol or diesel)
- age (less than 3 years or 3 or more years)
- air conditioning (with or without)

This Venn diagram has eight sections where numbers can be written. These match with the eight cells in the table.

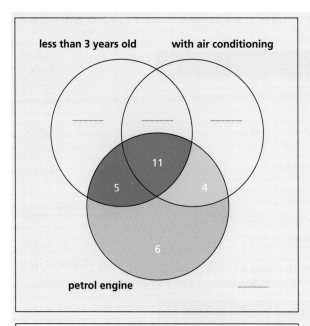

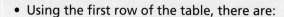

- Using the first row of the table, there are:
- 11 cars that have a petrol engine and air conditioning and are less than 3 years old
- 5 cars that have a petrol engine but no air conditioning and are less than 3 years old
- 4 cars that have a petrol engine and air conditioning and are 3 or more years old
- 6 cars that have a petrol engine but no air conditioning and are 3 or more years old.

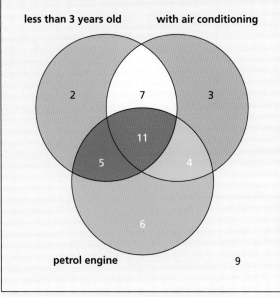

The numbers in the second row can be entered in a similar way.

Exercise 2

1 Eric wants to collect data about the heights of trees in a woodland. Eric knows that no tree is taller than 40 metres. Copy and complete this data collection sheet by writing in suitable class intervals. All class intervals should have the same width.

Height, h (m)	Tally	Total
....		
....		
....		
....		
....		

2 Channa works in a hospital. He designs this data collection sheet for recording the length (in cm) of the feet of newborn babies.

Length, x (cm)	Tally	Total
$0 \leq x \leq 10$		
$10 \leq x \leq 20$		
$20 \leq x \leq 25$		
$25 \leq x \leq 27$		
$27 \leq x \leq 30$		

a) Write down two problems with Channa's data collection sheet.

b) Design a better data collection sheet for Channa to use.

3 Gina designs a data collection sheet to record the number of cars passing her house per minute. She tests her data collection sheet by recording the number of cars for 15 one-minute intervals.

Number of cars	Tally	Total
Under 20		0
21 – 30		0
31 – 40	IIII	4
41 – 50	||||| I	6
Over 50	|||||	5

a) Explain why Gina should refine her data collection sheet.

b) Design a new data collection sheet that might be more suitable for Gina to use.

4 Mason measures the length (in cm) of some leaves:

11.4	6.3	9.8	13.2	8.5	16.3	5.4	7.9	10.2	11.5
8.6	7.0	6.6	8.7	12.1	9.9	8.7	10.7	8.5	11.2
14.8	17.2	12.6	10.4	8.7					

a) Mason first uses this frequency table to summarise his data.

Length, L (cm)	Tally	Frequency
$0 < L \leq 10$		
$10 < L \leq 20$		
$20 < L \leq 30$		
$30 < L \leq 40$		

Complete Mason's frequency table.

b) Write down a problem with Mason's frequency table.

c) Design and fill in a better frequency table for Mason's data.
Your classes should have equal widths.

5 Here are the ages, *A* (years), of 30 people in a cinema.

44	64	41	53	58	45	55	54	62	51
50	47	58	7	49	52	43	47	52	49
52	58	53	50	47	44	56	62	51	58

a) Show the data in a frequency table with these class intervals:

$0 \leq A < 20$, $20 \leq A < 40$, $40 \leq A < 60$ and $60 \leq A < 80$

b) Now show the data in a frequency table with these class intervals:

$A < 40$, $40 \leq A < 45$, $45 \leq A < 50$, $50 \leq A < 55$, $55 \leq A < 60$ and $60 \leq A < 65$

c) Which table best shows the data? Give a reason for your answer.

6 The table shows information about the paintings on show in an exhibition.

	Portrait		Not a portrait	
	Small size	Large size	Small size	Large size
Oil paint	8	16	7	32
Not oil paint	6	10	5	55

a) Find how many portraits there are on show in the exhibition.

b) Find the percentage of portraits that are small.

c) Copy and complete the Venn diagram to show this information.

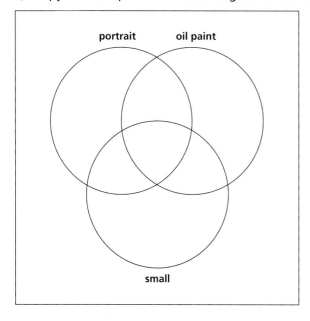

7 The incomplete table and the Venn diagram show some information about houses in a road.

	Front garden		No front garden	
	Red roof	Not red roof	Red roof	Not red roof
Black front door	4	8	7	
Not black front door				

a) Copy and complete the table and the Venn diagram.

b) Find the number of houses that do not have a red roof.

c) Simplify the ratio number of houses with a front garden : number of houses with no front garden.

d) Kendra says that fewer than 40% of houses on the road have a black front door. Is she correct? Give a reason for your answer.

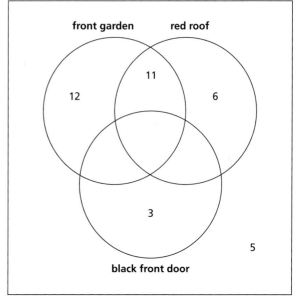

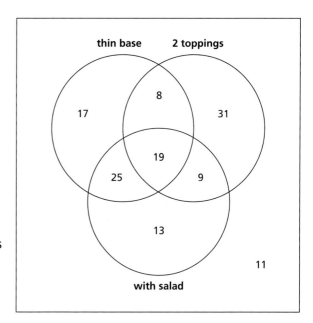

8 A pizza restaurant serves pizzas:

- with a thin base or stuffed crust
- with 2 toppings or with 3 toppings
- with salad or without salad.

The Venn diagram shows the pizzas ordered in one day.

a) Find how many customers ordered a pizza:
   i) with 2 toppings and with salad
   ii) with 3 toppings and with a thin base
   iii) with a stuffed crust.

b) Find whether more customers ordered pizzas with salad or without salad. Show your workings.

9 Alexa makes jewellery. One month she receives the following orders:

- 3 bracelets made of silver, to be engraved
- 7 pieces of jewellery made from silver, to be engraved
- 4 silver bracelets
- 14 engraved bracelets
- 10 silver pieces of jewellery
- 30 pieces of jewellery to be engraved
- 23 bracelets
- 25 pieces of jewellery that are not bracelets.

a) Copy and complete the Venn diagram to show the information.

b) Calculate the total number of orders she received that month.

c) What fraction of the orders does she not need to engrave? Give your answer in its simplest form.

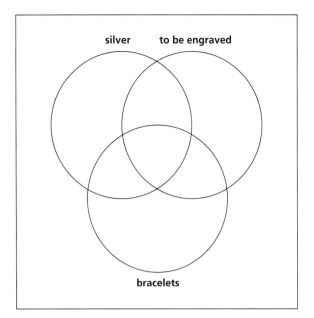

Thinking and working mathematically activity

The Venn diagram shows the items bought by customers visiting a farm shop one morning.

- Write some statements based on this Venn diagram. Some of your statements should be true and some should be false.

  For example,
  35 customers bought onions and carrots.
  (a statement which is FALSE).

- Swap your statements with a partner. Sort your partner's statements according to whether they are true or false.

- If you have time, create some true statements based on a Venn diagram you have made up. See if your partner can use your statements to discover what your Venn diagram must be.

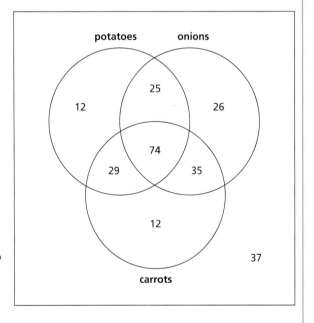

Consolidation exercise

1 Pablo runs an ice cream shop.

a) Here are some variables linked to the ice creams his customers order.

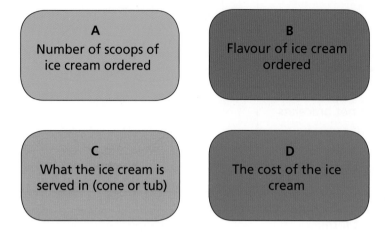

A
Number of scoops of ice cream ordered

B
Flavour of ice cream ordered

C
What the ice cream is served in (cone or tub)

D
The cost of the ice cream

Copy the table and write each variable in the correct column.

Qualitative data	Quantitative data

b) The Venn diagram shows some information about the ice creams Pablo sold one afternoon.

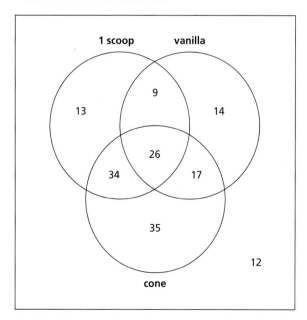

i) Find the number of customers who ordered more than one scoop of ice cream.

ii) Find the total number of ice creams sold that afternoon.

iii) Find the percentage of ice creams sold that were in a cone.

Iqbal wants to record the mass of plums grown on two trees. He expects all the plums to have a mass between 40 g and 70 g.

Copy and complete this data collection sheet so that it is suitable for recording Iqbal's data. All intervals should have the same width.

Mass, x (grams)	Tree 1	Tree 2
.... $\leq x <$		
.... $\leq x <$		
.... $\leq x <$		
.... $\leq x <$		
.... $\leq x <$		
.... $\leq x <$		

3 Jana is the manager of a corner shop. She wants to obtain a sample of people from the local area to find out how often they visit her shop.

She stands outside her shop and interviews a sample of people as they enter.

Give a reason why she may not get useful data.

4 Victor and Noor are investigating the area of postage stamps. They design these data collection sheets.

Victor

Area, A (cm²)	Tally
$0 < A \leq 2$	
$2 < A \leq 4$	
$4 < A \leq 6$	
$6 < A \leq 8$	
$8 < A \leq 10$	
$10 < A \leq 12$	

Noor

Area, A (cm²)	Tally
$0 - 10$	
$20 - 30$	
$40 - 50$	
$60 - 70$	
$80 - 90$	

Whose data collection sheet would be more suitable for recording the areas of stamps? Give reasons for your answer.

5 Jameela records the time (to the nearest second) that some runners took to complete a race.

54	57	55	59	52	53	58	59	61	60	55
57	59	60	57	58	54	58	57	58	61	54

She considers two different ways of grouping the data.

Method 1 uses these class intervals

$52 \leq t < 54$
$54 \leq t < 56$
$56 \leq t < 58$
$58 \leq t < 60$
$60 \leq t < 62$

Method 2 uses these class intervals

$51.5 \leq t < 53.5$
$53.5 \leq t < 55.5$
$55.5 \leq t < 57.5$
$57.5 \leq t < 59.5$
$59.5 \leq t < 61.5$

Tip

One of the runners took 54 seconds (to the nearest second). Think about how long that runner might have actually taken to run the race.

Why are the intervals given in Method 2 more suitable than the intervals in Method 1?

End of chapter reflection

You should know that...	You should be able to...	Such as...
Quantitative data is numerical and qualitative data is non-numerical.	Identify whether data are quantitative or qualitative.	Which of these are qualitative? • Shoe colour • Favourite drink • Volume
Bias can result when collecting data if some groups of people are over-represented or under-represented in a survey.	Discuss data collection methods and identify possible sources of bias.	Pinja wants to know the time that students in her town leave for school in the morning. She asks 30 children who live in her road. Give a reason why Pinja's data may not be reliable.
Data can be summarised in frequency tables, two-way tables and Venn diagrams.	Reflect critically on the appropriateness of a given frequency table.	Give a reason why this frequency table is not very appropriate. | Age (years)	Frequency
---	---
0 – 14	1
15 – 29	0
30 – 44	0
45 – 59	0
60 – 64	37	
	Interpret and record information in a Venn diagram.	Using this Venn diagram, find the fraction of all people who have brown hair and glasses.

Standard form

You will learn how to:
- Multiply and divide integers and decimals by 10 to the power of any positive or negative number.
- Understand the standard form for representing large and small numbers.

Starting point

- how to multiply or divide a whole number or a decimal by a positive integer power of 10?
  For example, find 0.4136×10^3 and $52 \div 10^4$
- how to multiply or divide a whole number or a decimal by 0.1 or 0.01?
  For example, find 72×0.1 and $6.91 \div 0.01$

- you learn to do calculations with numbers written in standard form.

4.0 Getting started

Higher or lower?

Choose any starting number between 0 and 20.

You are going to multiply your number by 0.1. Do you think your answer will be **higher** or **lower**?

Check your answer using a calculator. Did you predict correctly?

Next, divide your answer by 0.001. Before you divide, predict whether your answer will be **higher** or **lower**.

Continue for the rest of the chain.

My number → × 0.1 → ÷ 0.001 → ÷ 0.01 → × 0.0001 → × 0.001 → ÷ 0.001

Repeat using a new number. This time, start with a decimal number between 0 and 1.

Can you predict the answer before you calculate?

Which calculations in the chain make the number higher?

Which calculations make the number lower? Can you explain this?

Key terms

Powers of 10:

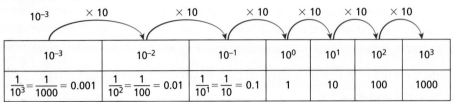

10^{-3}	10^{-2}	10^{-1}	10^0	10^1	10^2	10^3
$\frac{1}{10^3}=\frac{1}{1000}=0.001$	$\frac{1}{10^2}=\frac{1}{100}=0.01$	$\frac{1}{10^1}=\frac{1}{10}=0.1$	1	10	100	1000

Multiplying a number by 10^{-1}, 10^{-2}, 10^{-3} ... has the same effect as dividing the number by 10, 100, 1000...

Dividing a number by 10^{-1}, 10^{-2}, 10^{-3} ... has the same effect as multiplying the number by 10, 100, 1000...

Worked example 1

Work these out:

a) 0.036 × 10^4 **b)** 21 × 10^{-2} **c)** 14.3 ÷ 10^3 **d)** 0.02 ÷ 10^{-1}

> **Tip**
>
> The priority of operations is:
> Brackets, Indices or Powers, Multiplication and Division, Addition and Subtraction.

a) 0.036 × 10^4

= 0.036 × 10 000

= 360

Multiplying by 10^4 is the same as multiplying by 10 four times.

Each digit moves to the **left** four times.

100	10	1	•	0.1	0.01	0.001
		0	•	0	3	6
3	6	0	•			

b) 21 × 10^{-2}

= 21 × $\frac{1}{10^2}$

= 21 × $\frac{1}{100}$

= 21 ÷ 100 = 0.21

10^{-2} = $\frac{1}{10^2}$ = $\frac{1}{100}$ Multiplying by 10^{-2} is the same as dividing by 100.

Each digit moves to the **right** two times.

100	10	1	•	0.1	0.01	0.001
	2	1	•			
		0	•	2	1	

c) 14.3 ÷ 10^3

= 14.3 ÷ 1000

= 0.0143

Dividing by 10^3 is the same as dividing by 1000.

Each digit moves to the **right** three times.

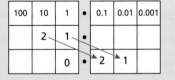

100	10	1	•	0.1	0.01	0.001	0.0001
	1	4	•	3			
		0	•	0	1	4	3

d) 0.02 ÷ 10^{-1}

= 0.02 ÷ $\frac{1}{10}$

= 0.02 × 10

= 0.2

10^{-1} = $\frac{1}{10^1}$ = $\frac{1}{10}$ Dividing by 10^{-1} is the same as multiplying by 10.

Each digit moves to the **left** one time.

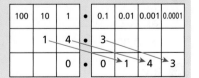
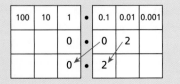

100	10	1	•	0.1	0.01	0.001
		0	•	0	2	
		0	•	2		

Thinking and working mathematically activity

Copy and complete the table showing a pattern of multiplications of 2.8 by different powers of 10.

Under each arrow, write the operation that changes one number into the next number.

	$\times\ 10^{-3}$		$\times\ 10^{-1}$			$\times\ 10^{2}$		
				2.8			2800	

Copy and complete the table showing a pattern of divisions of 2.8 by different powers of 10.

Under each arrow, write the operation that changes one number into the next number.

	$\div\ 10^{-3}$		$\div\ 10^{-1}$			$\div\ 10^{2}$		
			28	2.8			0.0028	

Which calculations give the answer 2.8?

Which types of calculation give answers greater than 2.8? Explain why.

Which types of calculation give answers less than 2.8? Explain why.

Exercise 1

1 Find:

a) 0.53×100 b) 260×0.1 c) 0.43×0.01 d) 897×0.001

e) $2089 \div 10$ f) $0.68 \div 100$ g) $3.79 \div 0.1$ h) $0.067 \div 0.01$

2 a) Write the number 10 000 as a power of 10.

b) Write the number 10^{-1} as a fraction and as a decimal.

c) Write the number $\dfrac{1}{1000}$ as a power of 10 and as a decimal.

d) Write the number 0.0001 as a power of 10 and as a fraction.

3 Find:

a) 13×10^{3} b) 0.806×10^{2} c) 0.0065×10^{4} d) 7.7×10^{5}

e) $315 \div 10^{2}$ f) $40.8 \div 10^{3}$ g) $0.69 \div 10^{1}$ h) $18 \div 10^{6}$

4 Find:

a) 202×10^{-2} b) 382.6×10^{-2} c) 0.52×10^{-1} d) 6521×10^{-4}

e) $0.056 \div 10^{-2}$ f) $569 \div 10^{-1}$ g) $0.1 \div 10^{-3}$ h) $75 \div 10^{-4}$

5 State whether or not each expression is equivalent to an integer.

a) 20×10^{-1} b) 8×10^{-2} c) 6.8×10^{-3} d) 0.51×10^{-4}

e) $3.06 \div 10^{-1}$ f) $0.045 \div 10^{-2}$ g) $51 \div 10^{-3}$ h) $0.039 \div 10^{-3}$

6 True or false? Explain your answer.

a) $8 \times 10^{-2} = -800$

b) $3 \times 10^{-1} = 0.3$

c) $0.16 \times 10^{-3} = 160$

d) $8.3 \div 10^{-3} = 83\,000$

e) $0.75 \div 10^{-1} = 7.5$

f) $0.039 \div 10^{-2} = 39$

7 Copy and complete.

a) $52 \times 10^{-3} = \ldots\ldots$

b) $28 \div 10^{-2} = \ldots\ldots$

c) $0.48 \div 10^{-4} = \ldots\ldots$

d) $\ldots\ldots \times 10^{4} = 3210$

e) $\ldots\ldots \times 10^{-2} = 0.79$

f) $\ldots\ldots \div 10^{-4} = 0.65$

8 Use < or > or = to make these sentences true.

a) $10^{-2} \ldots\ldots -100$

b) $10^{0} \ldots\ldots 0$

c) $3 \times 10^{-1} \ldots\ldots 30$

d) $0.067 \div 10^{-4} \ldots\ldots 1$

9 Change one number in each calculation to make it correct.

a) $0.45 \times 10^{-1} = 4.5 \div 10^{1}$

b) $8 \times 10^{3} = 0.8 \times 10^{2}$

c) $8 \times 10^{-2} = 0.8 \div 10^{3}$

d) $2300 \div 10^{3} = 2.3 \times 10$

e) Is there only one possible way to correct each calculation? Explain why.

4.2 Standard form

Key terms

Standard form, also called **scientific notation**, is a way of writing any number. Standard form looks like $a \times 10^{b}$, where $1 \le a < 10$ and b is a positive or negative integer. It is particularly helpful for writing very large and very small numbers in a shorter way.

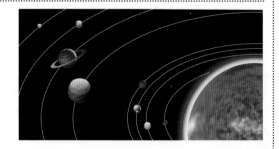

Worked example 2

a) Write 8.2×10^4 as an ordinary number.

b) Write 3.01×10^{-5} as an ordinary number.

c) Write 562 000 in standard form.

d) Write 0.0046 in standard form.

---

a) $8.2 \times 10^4 = 82\,000$

Multiplying by 10^4 is the same as multiplying by 10 four times, or multiplying by 10 000.

Move each digit four places to the left, and write zeros in the empty places.

10000	1000	100	10	1	•	0.1
				8	•	2
8	2	0	0	0	•	

---

b) $3.01 \times 10^{-5} = 0.0000301$

Multiplying by 10^{-5} is the same as dividing by 10 five times, or dividing by 100 000.

Move each digit five places to the right, and write zeros in the empty places.

1	•	0.1	0.01	0.001	0.0001	0.00001	0.000001	0.0000001
3	•	0	1					
0	•	0	0	0	0	3	0	1

---

c) $562\,000 = 5.62 \times 10^5$

Move the decimal point to make 5.62, which is greater than or equal to 1 and less than 10.

Find how many times to multiply or divide this number by 10 to make the original number.

$562\,000 = 5.62 \times 10^?$

100000	10000	1000	100	10	1	•	0.1	0.01
					5	•	6	2
5	6	2	0	0	0	•		

$562\,000 = 5.62 \times 10^5$

---

d) $0.0046 = 4.6 \times 10^{-3}$

Move the decimal point to make 4.6, which is greater than or equal to 1 and less than 10.

Find how many times to multiply or divide this number by 10 to make the original number.

$0.0046 = 4.6 \times 10^?$

1	•	0.1	0.01	0.001	0.0001
4	•	6			
0	•	0	0	4	6

$0.0046 = 4.6 \times 10^{-3}$

Exercise 2

1–6

 1 State whether each number is in standard form.

For each number that is not in standard form, explain how you know.

a) 4.5×10^{16} b) 0.27×10^{-8} c) 12×10^3

d) 1.00×10^{-7} e) 7.008×1^{-4} f) 10×10^9

2 Write in standard form:

a) $7 \times 100\,000$ b) 1.45×1000 c) $2.7 \div 10\,000$

d) $6 \div 1\,000\,000$ e) $3.41 \times 100\,000\,000$ f) $5.9 \div 1000$

3 Write as an ordinary number:

a) 6×10^5

b) 3.8×10^4

c) 7×10^{-6}

d) 8.03×10^{-3}

e) 1×10^6

f) 9.09×10^{-4}

g) 6.261×10^7

h) 5.077×10^{-5}

i) 9.2×10^3

4 Write in standard form:

a) 20 000

b) 0.000082

c) 356 000

d) 0.03801

e) 1 500 000

f) 560

g) 0.000571

h) 17

i) 0.1

5 Benigno, Melinda and Sonto want to write the number 1 in standard form.

Benigno says it is 1×10^1

Melinda says it is 1×10^0

Sonto says it is just 1

Explain which answer is correct. Explain why the other two answers are incorrect.

6 Write each number in standard form.

a) 50×10^3

b) 0.8×10^4

c) 306×10^5

d) 44×10^{-3}

e) 0.1×10^{-4}

f) 10×10^6

> **Tip**
>
> You may find it easier to convert to ordinary numbers first.

7 **Technology question** Most scientific calculators have a standard form mode, or 'Sci' mode. In this mode, all calculated numbers are shown in standard form. Find out how to put your calculator into this mode.

Use your calculator to convert these numbers to standard form.

a) 720 000

b) 0.0000081

c) 50 800 000

d) 0.00004105

e) 9 346 415

f) 0.0000004

> **Tip**
>
> In standard form mode on many calculators, if you type an ordinary number and press '=' or 'enter', the calculator displays the number in standard form.

8 **Technology question** To enter a number in standard form, such as 2.7×10^{-11}, into a calculator, either:

type 2.7×10, then the index key (which may be marked ^ or x^y), and then −11, or

type 2.7, then the key labelled '$\times 10^x$', 'EXP' or 'EE' (which means '$\times 10$ to the power'), and then −11.

Use a calculator to do these calculations.

a) $(3 \times 10^6) \times (2 \times 10^7)$

b) $(6.8 \times 10^7) \div (4 \times 10^{-12})$

c) $(5 \times 10^{14}) + (7 \times 10^{13})$

d) $(3.7 \times 10^{-10}) - (5.8 \times 10^{-11})$

> **Discuss**
>
> In question 8, in which calculation(s) do the brackets affect the result? Explain your answers.

Thinking and working mathematically activity

Copy and complete the table, showing the same length in millimetres, centimetres, metres and kilometres, as ordinary numbers and in standard form.

Quantity in ordinary form	Quantity in standard form
.... mm	.... mm
500 000 cm	.... cm
5000 m	.... m
.... km	.... km

Make a similar table for the quantity 46 grams, showing the same mass in milligrams (mg, thousandths of a gram), centigrams (cg, hundredths of a gram), and kilograms.

Write rules for converting a quantity in standard form:

- from milli- to centi-
- from kilo- to normal (for example, kilometres to metres)
- from milli- to kilo-
- from kilo- to centi-

Explain why each rule works.

Consolidation exercise

Which of the answers are greater than 10?

a) $57 \div 10^{-2}$ **b)** 93×10^{-2} **c)** 0.087×10^{-1} **d)** $0.58 \div 10^{4}$

e) 0.068×10^{-1} **f)** 0.79×10^{-3} **g)** $6.2 \div 10^{-3}$ **h)** $0.85 \div 10^{-2}$

2 Complete each statement:

a) $7 \times 10 \ldots\ldots\ldots = 70$ **b)** $0.7 \div 10^{-3} = \ldots\ldots$

c) $0.07 \div 10^{-4} = 7 \times \ldots\ldots\ldots$ **d)** $700 \div 10^{3} = 0.007 \ldots\ldots \quad \ldots\ldots$

e) $70 \div 10^{-2} = \ldots\ldots \times 10^{-2}$ **f)** $700 \times \ldots\ldots = 0.007 \ldots\ldots \quad \ldots\ldots$

3 Copy and complete these multiplication grids.

×	10^5	10^{-3}
0.48		
	79	

×	10^{-3}	$10^{\square}$
	175	
8.6		0.086

×	$10^{\square}$	$10^{\square}$
3400		3.4
0.23	0.0023	

Write each of these quantities as ordinary numbers. (Do not change the units.)

a) The diameter of a human red blood cell is about 7×10^{-6} m.

b) The temperature at the centre of the Sun is 1.5×10^{7} °C.

c) The distance from the Earth to the Moon is 3.844×10^{5} m.

5 Write each of these quantities in standard form. (Do not change the units.)

a) The speed of light is 1 080 000 000 kilometres per hour.

b) The virus that causes influenza ('flu') has a diameter of approximately 0.0000001 m.

c) The distance from the Earth to the Sun is 149 million kilometres.

6 Write <, > or = between each pair of numbers to make correct statements.

a) 5×10^8 .... 8×10^8

b) 7×10^{-12} .... 1×10^{-12}

c) 3×10^{17} .... 3×10^{19}

d) 8×10^{-16} .... 8×10^{-17}

End of chapter reflection

You should know that...	You should be able to...	Such as...
If x is positive, then: • multiplying by 10^x is equivalent to multiplying by 10 x times • dividing by 10^x is equivalent to dividing by 10 x times • multiplying by 10^{-x} is equivalent to dividing by 10 x times • dividing by 10^{-x} is equivalent to multiplying by 10 x times.	Multiply and divide integers and decimals by integer powers of 10.	Find: a) 6.35×10^4 b) 720×10^{-6} c) $13 \div 10^7$ d) $0.88 \div 10^{-5}$
Any number can be written in standard form. Standard form is $a \times 10^b$, where $1 \le a < 10$ and b is an integer.	Convert between ordinary numbers and standard form.	Write in standard form: a) 0.00036 b) 5 904 000 Write as ordinary numbers: a) 3.5×10^5 b) 1.022×10^{-6}

5 Expressions

You will learn how to:
- Understand that the laws of arithmetic and order of operations apply to algebraic terms and expressions (four operations and integer powers).
- Understand how to manipulate algebraic expressions including:
  - expanding the product of two algebraic expressions
  - applying the laws of indices
  - simplifying algebraic expressions.
- Understand that a situation can be represented either in words or as an algebraic expression, and move between the two representations (including squares, cubes and roots).

Starting point

Do you remember...

- how to use index notation?

  For example, write $t \times t \times t$ using index notation.

- how to construct algebraic expressions from a given situation?

  For example, Eric has d. He spends $\frac{1}{2}$ on a holiday and buys b books costing m each to read while on holiday. Write an expression for the amount of money he has left.

- how to use the rules of BIDMAS to simplify algebraic expressions?

  For example, expand and simplify $3y(3y + 2) + 2(y - 7)$.

- how to factorise algebraic expressions by using the highest common factor of the terms and brackets?

  For example, write $12x^2 - 9xy$ as $3x(4x - 3y)$.

- how to substitute values into expressions to get an answer using the rules of BIDMAS?

  For example, find the value of $4x^2 + 7x - 2$ when $x = 3$.

This will also be helpful when...

- you learn to solve more complex equations
- you learn to change the subject or rearrange formulae
- you learn to use the quadratic formula to solve quadratic equations.

5.0 Getting started

Substitution race track

This is a game to play in pairs. You will need a dice, two counters and a copy of the race track.

START/FINISH $6 - n$	$n - 4$	$4n - 1$	$n^2 - 9$	$2n + 6$	$2 - n^2$
$2n^2 - 3n$					$3n$
$3n$					$3n^2 - 4n$
$n^2 - 10$					$5 - n$
$n - n^2$					$n^2 - n$
$3 - n$					$9 - n^2$
$2n^2 - n - 6$					$n^2 - 4$
$n^2 - 12$	$7 - n$	$8 - 3n$	$2n$	n^2	$7 - n^2$

How to play

Each player places their counter on the START/FINISH space.

Roll the dice.

Substitute the number on the dice into the expression on the space to work out how many spaces to move your counter.

Move forward/backward according to the answer calculated.

For example, say you are on the space marked $n^2 - 4$. If you roll a 3 you would move five spaces forward, but if you roll a 1 you would move three spaces backward.

Take turns rolling the dice and moving your counter. The winner is the first person to go around the board three times.

Ready, set, go!

5.1 Substitution

Worked example 1

Using the formula $s = ut + \frac{1}{2}at^2$, find the value of s, when $u = 7.5$, $t = 25$ and $a = -0.2$

$s = 7.5 \times 25 + \frac{1}{2} \times (-0.2) \times 25^2$	You must use the correct order of operations, i.e., you must square the value of t first.
$= 7.5 \times 25 + \frac{1}{2} \times (-0.2) \times 625$	You then multiply the parts separately before adding or subtracting at the end.
$= 187.5 + (-62.5)$ $= 125$	You must also apply your knowledge of addition and subtraction of negative numbers.

Tip

Remember the rules for addition, subtraction, division and multiplication with directed numbers.

1 When $x = -4$ and $y = 0.1$, find the value of:

   a) $4xy$
   b) $(x + y)^2$
   c) $y(x - 2)$
   d) $x^2 - 2y$
   e) $xy + 9$

   f) x^2y
   g) $3(x + y)$
   h) $7y - x^2$
   i) $(3x - y)^2$

2 When $a = 0.5$, $b = 6$, $c = -9$, $d = -1.2$ and $e = -2$, find the value of:

   a) ab
   b) $4a - 2$
   c) $2a - c$
   d) $b^2 + c$
   e) $bc + e$

   f) bd
   g) $c^2 - e$
   h) $\dfrac{d}{e}$
   i) $\dfrac{e^2}{a}$

3 When $a = 9$, $b = 4$, $c = -\dfrac{1}{5}$, $d = -5$ and $e = -1$, find the value of:

   a) b^3
   b) e^3
   c) $\sqrt{a}$
   d) $bc + e$
   e) $3(a - d)$

   f) $b(d - e)$
   g) bd^2
   h) $(d - e)^2$
   i) $(3 + d)(4 - e)$

4 Using the formula $v^2 = u^2 - 2as$, which of these values of v, u, a and s can be substituted to satisfy the equation?

   a) $v = 5$, $u = 4$, $a = 0.5$ and $s = 9$
   b) $v = 4$, $u = 5$, $a = 0.5$ and $s = -9$

   c) $v = 5$, $u = 4$, $a = -0.5$ and $s = -9$
   d) $v = 5$, $u = 4$, $a = 0.5$ and $s = -9$

   Explain your answer.

5 When $m = 4$, $n = 3$, and $p = -2$, find the value of:

   a) $3(2m - n)^4$
   b) $(m + 2n)^3$
   c) $(m + np)^3 + mnp^4$
   d) $\dfrac{m^3 - 4p}{n^2 - 3}$

6 When $x = 2$, $y = 12$ and $z = 15$, find the value of:

   a) $\dfrac{120}{5x^2 - 8}$
   b) $\sqrt{\dfrac{2x}{y + 4}}$
   c) $12x^3 - \dfrac{5y}{z}$
   d) $\dfrac{1}{8}x^3 + 2xy - 4$

7 When $a = 3$, $b = -4$ and $c = 5$, have these substitutions been correctly calculated? If they have not, show what the correct calculation and answer should be.

   a) $(5a - 2c)^3 + b^2 = 109$
   b) $\dfrac{4c^3 - 3ab}{4} = 24$

Thinking and working mathematically activity

You have two dice. Roll one, then the other.

If the number on the first dice is even, the number on the second dice is positive. Write it down.

If the number on the first dice is odd, the number on the second dice is negative. Write it down.

Repeat this three times. Label the numbers a, b and c, in order.

Substitute your values into these expressions. Which is the largest? Are there any values that you cannot calculate?

$$(3a - 2b)(c + 4)^2 \qquad \sqrt{a^3 + 2bc} \qquad \dfrac{4b^3 - ac}{2b}$$

Roll the dice again to get three more numbers. What happens to your expressions now?

Try to find the set of numbers that will give the largest value for each expression.

5.2 Expanding a pair of brackets

Key terms

When you **simplify** an algebraic expression, you apply the rules of arithmetic to the expression to write it in a shorter or simpler form.

When you **expand brackets**, you multiply out the terms in brackets, then simplify your answer by adding or subtracting like terms.

A **quadratic expression** is an expression that includes an unknown whose highest power is 2. These are all examples of quadratic expressions:

$x^2 + 2x + 1$ $2g^2 + 6g$ $p^2 - 9$ $25 - y^2$

Worked example 2

Expand and simplify the expression $(x + 4)(x - 5)$.

$(x + 4)(x - 5)$	Multiply each term in the first pair of brackets by each term in the second pair of brackets. You can use lines to connect terms which you need to multiply: You can also use the grid method. Simplify your answer.	
$= x^2 + 4x - 5x - 20$ $= x^2 - x - 20$		

Exercise 2

1 Expand and simplify each expression:

a) $(x + 2)(x + 1)$ b) $(x + 3)(x + 2)$ c) $(x + 1)(x + 1)$ d) $(x + 12)(x + 1)$

e) $(x + 3)(x + 4)$ f) $(x + 2)(x + 6)$ g) $(s + 1)(s + 15)$ h) $(t + 4)(t + 2)$

i) $(v + 8)(v + 1)$ j) $(w + 4)(w + 5)$ k) $(y + 7)(y + 2)$ l) $(z + 2)(z + 2)$

2 Expand and simplify each expression:

a) $(x - 1)(x + 2)$ b) $(x - 4)(x + 3)$

c) $(k - 7)(k + 2)$ d) $(m + 5)(m - 2)$

e) $(n + 8)(n - 1)$ f) $(p + 6)(p - 3)$

g) $(q - 2)(q - 10)$ h) $(r - 6)(r - 2)$

i) $(s - 1)(s - 9)$ j) $(t - 3)(t + 4)$

k) $(x + 4)(x - 2)$ l) $(x - 4)(x + 2)$

Tip

Remember what happens when you multiply positive and negative numbers:

positive × positive = positive

positive × negative = negative

negative × positive = negative

negative × negative = positive

Thinking and working mathematically activity

Expand and simplify $(x + 4)(x - 4)$.

What do you notice? Why do you think this happens?

Try to create other pairs of brackets like this.

Can you make a general rule for expanding brackets in this form?

Now expand and simplify $(x + 4)^2$.

Try to create other pairs of brackets like this.

Can you make a general rule for expanding brackets in this form?

3 Abdullah expands $(x + 3)^2$ and gets the answer $x^2 + 9$.

    a) Describe what he has done wrong.

    b) Expand $(x + 3)^2$ correctly and simplify.

> **Tip**
>
> Remember:
> $(x + 4)^2 = (x + 4)(x + 4)$

4 Expand and simplify

    a) $(x + 5)^2$ **b)** $(n + 10)^2$ **c)** $(y - 6)^2$ **d)** $(p - 11)^2$

    e) $(x + 7)(x - 7)$ **f)** $(x - 1)(x + 1)$ **g)** $(m + 9)(m - 9)$ **h)** $(r - 14)(r + 14)$

5 Complete these statements.

    a) $(x + 1)(x + ...) = x^2 + 4x + 3$ **b)** $(x + 4)(x + ...) = x^2 + 6x + 8$

    c) $(x - 3)(x + ...) = x^2 + 2x - 15$ **d)** $(x - 6)(x - ...) = x^2 - ... + 24$

    e) $(x + ...)(x - 2) = x^2 + ... x - 14$ **f)** $(x + ...)(... - 4) = x^2 - ... x - 4$

> **Tip**
>
> You may find it helpful to use a grid (like the one shown in Worked example 1).

6 **Vocabulary question** Copy and complete the sentences below with words in the box. You may use a word more than once.

term	brackets	four	expand	simplify	multiply

When you .... two brackets containing linear expressions, you must .... each

.... in the first pair of .... by each .... in the second.

You will get .... terms. Then you should .... if possible.

5.3 Laws of indices

Key terms

To multiply terms with indices, you use the **index law for multiplication**, which states that $a^m \times a^n = a^{m+n}$.

To divide terms with indices, you use the **index law for division**, which states that $a^m \div a^n = a^{m-n}$.

Worked example 3

Simplify:

a) $m \times n \times n \times m \times n \times n \times n$ b) $w^2 \times w^3$ c) $p^6 \div p^2$

d) $6z^7 \div 3z^2$ e) $(a^4)^3$

a) $m \times n \times n \times m \times$ $n \times n \times n$ $= m \times m \times n \times n \times$ $n \times n \times n$ $= m^2 n^5$	**Tip** It might help you to rewrite the statement with all of the ms together and all of the ns together. Simplify the product of the ms using powers: $m \times m = m^2$ Simplify the product of the ns using powers: $n \times n \times n \times n \times n = n^5$ Combine to get $m^2 n^5$	$m \times n \times n \times m \times n \times n \times n$ $= m \times m \times n \times n \times n \times n \times n$ $= m^2 \times n^5$
b) $w^2 \times w^3$ $= w^{2+3}$ $= w^5$	Use the index law for multiplication: add the powers.	$w \times w \ \times \ w \times w \times w = w \times w \times w \times w \times w$ $w^2 \ \times \quad w^3 \quad = \quad\quad w^5$
c) $p^6 \div p^2$ $= p^{6-2}$ $= p^4$	Use the index law for division: subtract the powers.	$\dfrac{p \times p \times p \times p \times p \times p}{p \times p} = p \times p \times p \times p$ $\dfrac{p^6}{p^2} = p^4$
d) $6z^7 \div 3z^2$ $= (6 \div 3) \times z^{7-2}$ $= 2z^5$	Divide the numbers and use the index law for the powers.	$\dfrac{6 \times z \times z \times z \times z \times z \times z \times z}{3 \times z \times z}$ $= \dfrac{6}{3} \times \dfrac{z \times z \times z \times z \times z \times z \times z}{z \times z}$ $= 2 \times z^5$ $= 2z^5$
e) $(a^4)^3$ $= a^4 \times a^4 \times a^4$ $= a^{12}$	To the power 3 means you multiply whatever is inside the brackets by itself 3 times. Then use the index law for multiplication.	$(a^4)^3 = a^4 \times a^4 \times a^4$ $= a \times a \times a \times a \times$ $a \times a \times a \times a \times a \times a \times a \times a$ $= a^{12}$

Discuss

Look at part **e)** above. Can you see a connection between the starting powers and the answer? Does this work for other powers written in the form $(a^m)^n$?

1 **Technology question** Use the internet to research the mathematician Al'Khwarizmi. Find out about his life and work and why he is sometimes known as 'the father of algebra'.

> **Did you know?**
>
> The word 'algebra' comes from the word, Al-jabr, which was a word used by Al'Khwarizmi in his work.

2 Use index notation to simplify these expressions:

a) $n \times n \times n \times n \times n \times n$

b) $h \times h \times h \times h \times h$

c) $p \times p \times p \times p \times p \times p \times p$

d) $6 \times e \times e \times 2 \times e$

e) $t \times u \times u \times t \times t \times t$

f) $a \times c \times b \times b \times a \times b \times a$

g) $5 \times y \times 3 \times y \times 2 \times y \times y$

h) $q \times 4 \times q \times r \times q \times 9$

3 Simplify:

a) $a^4 \times a^3$

b) $c^6 \times c^2$

c) $g^3 \times g^2$

d) $p^3 \times p^5$

e) $t^6 \times t^9$

f) $u^4 \times u^2 \times u^3$

4 Decide whether each of these expressions has been simplified correctly. If the simplification is wrong, say what the correct answer should be.

a) $k^9 \div k^3 = k^3$

b) $y^7 \div y^2 = y^5$

c) $n^{11} \div n^5 = n^6$

d) $t^8 \div t = 8$

e) $r^5 \div r = r^5$

f) $\dfrac{p^4 \times p^8}{p^3} = p^9$

5 Simplify each expression:

a) $g^7 \div g^2$

b) $z^{12} \div z^4$

c) $\dfrac{h^9}{h^3}$

d) $\dfrac{f^{11}}{f^{10}}$

e) $\dfrac{y^{10}}{y^4} \times y$

f) $\dfrac{r^3 \times r^8}{r^7}$

> **Tip**
>
> Remember that $y = y^1$

6 Sven says that $a^5 \times b^2$ simplifies to ab^7. Explain why Sven is wrong.

7 Find the missing expressions:

a) $t^4 \times \boxed{} = t^9$

b) $u^3 \times \boxed{} \times u^7 = u^{11}$

c) $y^5 \times y^2 = y^3 \times \boxed{}$

d) $\boxed{} \div a^4 = a^{10}$

e) $\dfrac{q^{12}}{\boxed{}} = q^6$

f) $\dfrac{\boxed{} \times m^5}{m^3} = m^6$

g) $\dfrac{e^{13}}{e \times \boxed{}} = e^7$

h) $a^3 b^2 \times \boxed{} = a^5 b^9$

i) $2p^2 q^6 \times \boxed{} = 6p^3 q^8$

8 Match the expressions:

$(a^3)^5$	$2a^4$
$(2ab)^3$	$24a^9$
$12a^8 \div 3a^3$	a^{15}
$2(ab)^3$	$12a^3b^2$
$(a^2)^2 \times a \times a^3$	$2a^3b^3$
$4a^5 \times 6a^4$	a^8
$5a^4 - 3a^4$	$4a^5$
$2a^3 \times 6b^2$	$8a^3b^3$

9 Madina thinks that $d^4 + 2d^4 + 3d^4$ simplifies to $6d^{12}$.

Anchita thinks that it simplifies to $6d^4$.

Who is correct? Explain how you know.

Thinking and working mathematically activity

If $g^6 \times g^n = \dfrac{g^{12}}{g^m}$, find all possible values of n and m if they are positive integers.

What do you notice about the pairs of values for m and n?
Is this true only for 6 and 12?
Make up another similar statement, replacing 6 and 12 with other numbers.
What do you find about m and n now?
Explain to a partner what you have found out.

5.4 Algebraic fractions

Key terms

To manipulate **algebraic fractions**, you apply the same rules as you use for arithmetic fractions.

Worked example 4

Simplify:

a) $\dfrac{5h}{8} + \dfrac{h}{4}$

b) $\dfrac{12d - 3}{3}$

c) $\dfrac{4x^2 + 12xy}{x}$

d) $\dfrac{2y}{9} \times \dfrac{5y}{6}$

a) $\frac{5h}{8} + \frac{h}{4}$ $= \frac{5h}{8} + \frac{2h}{8}$ $= \frac{7h}{8}$	First find the common denominator. Use it to make equivalent fractions. Add the numerators to give a single fraction.	$\boxed{\frac{h}{8}}\ \boxed{\frac{h}{8}}\ \boxed{\frac{h}{8}}\ \boxed{\frac{h}{8}}\ \boxed{\frac{h}{8}}\ \boxed{\frac{h}{4}}$ $\boxed{\frac{h}{8}}\ \boxed{\frac{h}{8}}\ \boxed{\frac{h}{8}}\ \boxed{\frac{h}{8}}\ \boxed{\frac{h}{8}}\ \boxed{\frac{h}{8}}\ \boxed{\frac{h}{8}}$
b) $\frac{12d - 3}{3}$ $= \frac{3(4d - 1)}{3}$ $= 4d - 1$	Factorise the numerator. As you are multiplying by 3 and then dividing by 3, these operations cancel each other out, leaving you with $4d - 1$	$12d = 3 \times 4d$ $3 = 3 \times 1$ $\dfrac{3 \times (4d - 1)}{3}$
c) $\frac{4x^2 + 12xy}{x}$ $= \frac{x(4x + 12y)}{x}$ $= 4x + 12y$	Factorise x in the numerator, as that is what you have in the denominator. Multiplying by x and dividing by x cancel each other out, so you can cancel even though you do not have a value for x.	$4x^2 = 4 \times x \times x$ $12xy = 12 \times x \times y$ $\dfrac{x \times (4x + 12y)}{x}$ **Tip** When dividing by x, we must assume that x cannot be zero, because dividing by zero is not allowed.
d) $\frac{2y}{9} \times \frac{5y}{6}$ $= \frac{2y \times 5y}{9 \times 6}$ $= \frac{10y^2}{54}$ $= \frac{5y^2}{27}$	Just like with numerical fractions, you multiply the numerators and multiply the denominators. Simplify your answer if you can.	

Exercise 4

1 Write these as a single fraction.

   a) $\frac{h}{3} + \frac{7h}{12}$ **b)** $\frac{7h}{12} - \frac{h}{4}$ **c)** $\frac{a}{3} + \frac{a}{4} + \frac{a}{6}$ **d)** $\frac{7b}{8} - \frac{b}{4}$

2 Write these as a single fraction:

   a) $\frac{2}{x} + \frac{3}{x}$ **b)** $\frac{3}{y} - \frac{2}{y}$ **c)** $\frac{4}{5b} - \frac{2}{5b}$ **d)** $\frac{x}{3a} + \frac{y}{3a}$ **e)** $\frac{3}{2x} - \frac{1}{2x}$

3 In the sequences below, each term after the second term is formed by adding the two previous terms. Find the next three terms in each sequence.

   a) $\frac{x}{3}, \frac{y}{3}, \ldots, \ldots, \ldots$ **b)** $\frac{2}{a}, \frac{3}{a}, \ldots, \ldots, \ldots$ **c)** $\frac{c}{2}, \frac{d}{5}, \ldots, \ldots, \ldots$

4 Adam says that $\frac{3}{a} + \frac{3}{a} = \frac{6}{2a}$. Is he correct. Explain your answer.

5 Johan says that $\frac{3}{2x} + \frac{5}{6x} = \frac{11}{6x}$. Is he correct. Explain your answer.

> **Tip**
> Write the fractions so that both have denominators of $6x$.

6 Copy and complete these calculations:

a) $\dfrac{8n + 4}{4} = \dfrac{4(\boxed{} + 1)}{4} = \boxed{}$

b) $\dfrac{3p - 9}{3} = \dfrac{\boxed{}(p - \boxed{})}{3} = \boxed{}$

c) $\dfrac{12m + 18n}{6} = \dfrac{6(\boxed{} + \boxed{})}{6} = \boxed{}$

7 Simplify these fractions:

a) $\dfrac{6x + 2}{2}$ b) $\dfrac{21m - 7}{7}$ c) $\dfrac{12n + 8}{4}$ d) $\dfrac{10x^2 - 5x}{5}$

e) $\dfrac{6n + 18}{3}$ f) $\dfrac{m^2 + 3m}{m}$ g) $\dfrac{4a + 2ab}{a}$ h) $\dfrac{10x^2 - 5x}{x}$

8 Multiply these fractions. Simplify your answer when possible.

a) $\dfrac{3x}{4} \times \dfrac{x}{5}$ b) $\dfrac{4a}{5} \times \dfrac{3a}{8}$ c) $\dfrac{6y}{5} \times \dfrac{2}{3}$ d) $\dfrac{12b}{25} \times \dfrac{5b}{6}$

9 Here is Jon's homework on algebraic fractions.

He has got some of the questions right and some are wrong.

Identify any mistakes in these calculations and write the correct solutions.

a.	$\dfrac{12x - 3y}{3}$	$=$	$\dfrac{3(4x - y)}{3}$	$= 4x - y$
b.	$\dfrac{10x + 5}{5}$	$=$	$10x$	
c.	$\dfrac{12x + 4y}{4}$	$=$	$\dfrac{4(3x + y)}{4}$	$= 3xy$
d.	$\dfrac{6x^2 + 2x}{x}$	$=$	$\dfrac{2x(3x + 1)}{x}$	$= x(3x + 1)$
e.	$\dfrac{10m - 2m^2}{m}$	$=$	$\dfrac{2m(5 - m)}{m}$	$= 2(5 - m)$
f.	$\dfrac{3x^2 - x}{x}$	$=$	$\dfrac{3x(x - 1)}{x}$	$= 3(x - 1)$

10 Simplify these fractions by first factorising the numerators:

a) $\dfrac{6x + 3}{3} + \dfrac{4 - 2x}{2}$ b) $\dfrac{12a - 8}{4} + \dfrac{15a - 10}{5}$ c) $\dfrac{6x + 9y}{12} + \dfrac{2x + 8y}{4}$

11 Write this expression as a single fraction: $\dfrac{x + 3}{7} + \dfrac{x + 1}{5}$

> Tip
>
> Remember: when you multiply $x + 3$ by 5, this is the same as $5(x + 3) = 5x + 15$.

Working with a partner, make up a set of domino cards for algebraic fractions.

Remember that a question on one card has to match up with the answer on the next card.

The cards should all form a loop.

You should have at least 12 questions and answers.

Swap your cards with another pair to challenge them to solve your puzzle.

$$\frac{12d - 3}{3} \qquad 4d - 1$$

5.5 Constructing expressions

Worked example 5

a) Greta bakes n biscuits. She sells 8 biscuits for $2c$ cents each. She sells the remaining biscuits for c cents each. Form an expression for the total amount of money Greta makes.

b) I think of a number, n, multiply it by 3, square it, and then subtract 12. Write an expression for my number.

c) The area of a triangle is $\frac{1}{2}$ × base × height.

   Write an expression for the area of a triangle which has base $3x + 4$ and height $2x - 1$.

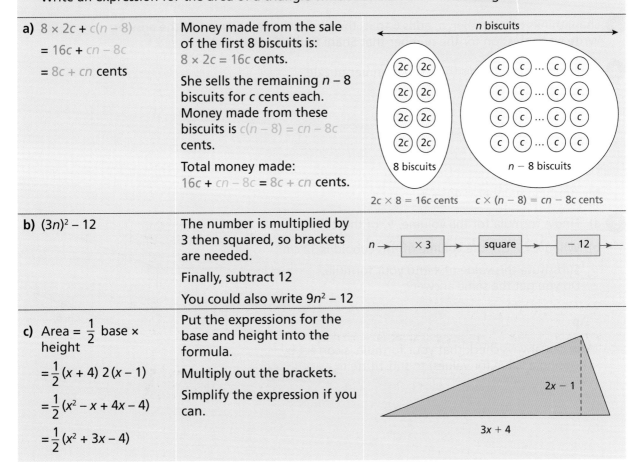

a) $8 \times 2c + c(n - 8)$ $= 16c + cn - 8c$ $= 8c + cn$ cents	Money made from the sale of the first 8 biscuits is: $8 \times 2c = 16c$ cents. She sells the remaining $n - 8$ biscuits for c cents each. Money made from these biscuits is $c(n - 8) = cn - 8c$ cents. Total money made: $16c + cn - 8c = 8c + cn$ cents.	
b) $(3n)^2 - 12$	The number is multiplied by 3 then squared, so brackets are needed. Finally, subtract 12 You could also write $9n^2 - 12$	
c) Area $= \frac{1}{2}$ base × height $= \frac{1}{2}(x + 4)\, 2\,(x - 1)$ $= \frac{1}{2}(x^2 - x + 4x - 4)$ $= \frac{1}{2}(x^2 + 3x - 4)$	Put the expressions for the base and height into the formula. Multiply out the brackets. Simplify the expression if you can.	

1 Here are some 'Think of a number' puzzles.

a) I think of a number, *n*.

I double it.

I subtract the answer from 30.

Find an expression for the answer.

b) I think of a number, *m*.

I add 4.

I multiply the answer by 3.

I then divide the answer by 5.

Find an expression for the answer.

2 Make up your own 'Think of a number' puzzles that match these expressions for the final answers:

	Starting number	Final answer
a)	*p*	$\dfrac{5p - 2}{4}$
b)	*q*	$\dfrac{4(6 - q)}{7}$
c)	*r*	$70 - \dfrac{2r}{9}$

3 Shani thinks of a number, *n*, adds 6 to it, then finds the square root of the answer. Write an expression for the number that Shani finishes with.

4 Copy and complete this think of a number puzzle:

I think of a number, *n*

I ....

I then ....

Finally, I ....

My finishing number is $4n^2 + 7$.

5 a) Find a formula for the volume, *V*, of this cuboid.

b) When *x* = 2 cm, the volume of this cuboid is 32 cm³.

Substitute this value of *x* into your formula. Do you get the same answer?

> **Tip**
>
> You can always check that your formula is correct by choosing a value (or values) to substitute in.

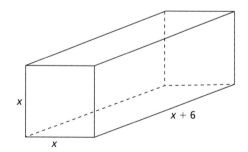

x

x

x + 6

6 Find expressions for each shaded area. All sides are measured in centimetres.

a)

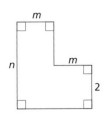

b)

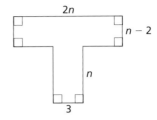

c)

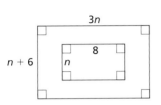

d)

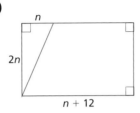

7 Draw compound shapes with areas given by these expressions.

a) $2n^2 + mn$ b) $\dfrac{n^2}{2} + 4n$ c) $n^2 + 4n + 8$

8 Let $m = 3$ and $n = 4$.

a) Write the dimensions of your shapes in question 7 and use these dimensions to calculate the area of each shape.

b) Substitute these values into each of the expressions in question 7 to check your answers to part a).

> **Tip**
>
> For 7c), try forming a shape from two rectangles. Make the length of one of the rectangles equal to $n + 4$.

9 The triangle and the square have the same area.

Find an expression, in terms of a and b, for the length of the side of the square.

Thinking and working mathematically activity

Think about different straight-edged shapes.
Draw diagrams of shapes that could have an area of $(12x^2 + 4)$ cm^2
Show the dimensions of your shapes clearly on each diagram.
Choose a value for x and check that your dimensions will give you the correct area.

▶ When $a = -4$, $b = 0.7$ and $c = 12$, find the value of the following expressions:

a) $3a + 4b - c$ b) $c^2 + 5ab$ c) $\sqrt{a^2 + 4c}$ d) $\dfrac{30a^2b}{c}$

▶ Using the formula $m = \dfrac{3\sqrt{n}}{2} + pq$, match the values of m with the correct values of n, p and q.

| $m = 1$ | $m = 5$ | $m = -1$ | $m = 7.5$ |

$n = 1$	$n = 4$	$n = 36$	$n = 100$
$p = 8$	$p = -0.8$	$p = -0.5$	$p = 7$
$q = 0.75$	$q = 5$	$q = 8$	$q = -2$

3 Expand and simplify each expression:

a) $(a + 5)(a + 2)$ b) $(b - 4)(b - 6)$ c) $(c + 5)(c - 7)$ d) $(d + 9)(d - 5)$

e) $(e + 5)^2$ f) $(f - 4)(f + 4)$ g) $(u - 9)(u + 8)$ h) $(x + 12)(x - 12)$

4 Match the equivalent expressions:

$(x + 3)(x - 4)$ $(x - 6)(x - 2)$ $(x + 12)(x + 1)$ $(x - 2)(x + 6)$

$x^2 + 13x + 12$ $x^2 - x - 12$ $x^2 + 4x - 12$ $x^2 - 8x + 12$

5 Write each expression as a single power of x:

a) $x^5 \times x^3$ b) $x^4 \times x^3 \times x^2 \times x$ c) $3x^4 \times 4x^3$ d) $x^{10} \div x^5$ e) $x^6 \div x$ f) $12x^7 \div 3x^2$

6 Emma thinks that $(3b^2)^3 \times 2b$ is $18b^7$.

Victor thinks that the answer is $54b^6$.

Vyan thinks it is $54b^7$.

Who is correct? Explain how you know.

Suggest what mistakes the others might have made.

▶ Simplify these fractions:

a) $\dfrac{6a + 12}{3}$ b) $\dfrac{15 - 20b}{5}$ c) $\dfrac{x^2 + 4x}{x}$ d) $\dfrac{2f + 6ef}{f}$ e) $\dfrac{3a}{4} \times \dfrac{a}{5}$ f) $\dfrac{3a}{4} \times \dfrac{2a}{9}$

8 Ellis buys p plants.

n of the plants each cost $\$d$.

The rest of the plants each cost $\$e$.

Write an expression for the total cost of the plants.

9 Polina thinks of a number, n, squares it, adds 6, then subtracts five times the number she first thought of. Which of these expressions give the same outcome?

$n^2 - 5n + 6$ $n^2 + 6 + 5n$ $n^2 + 6 - 5n$ $\sqrt{n} + 6 - 5n$

$(n - 3)(n - 2)$ $2n + 6 - 5n$ $(n + 2)(n + 3)$

10 The diagram shows shape $ABCDE$ formed from a rectangle and a triangle.

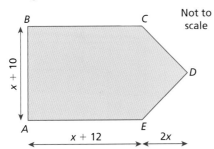

Not to scale

Write and fully simplify an expression for the area of the shape.

11 A shape has an area of $(x^2 + 5x + 6)$ cm^2.

Draw two possible shapes with this area. Mark the dimensions clearly on each diagram.

End of chapter reflection

You should know that...	You should be able to...	Such as...
When substituting a value into an expression or a formula, you must always apply the correct order of operations.	Substitute both positive and negative numbers into expressions and formulae.	Find the value of $3u + at^2$ when $u = 8$, $a = -2$ and $t = 3$
When you multiply out two sets of brackets, you multiply everything in the first pair of brackets by everything in the second pair, then simplify if possible.	Multiply out two sets of brackets and simplify your answer.	Multiply out and simplify: **a)** $(x - 3)(x + 4)$ **b)** $(t + 3)^2$ **c)** $(h - 9)(h + 9)$
The index law for multiplication states that $a^m \times a^n = a^{m+n}$	Use the index law for multiplication to simplify expressions.	Simplify $p^4 \times p^5$
The index law for division states that $a^m \div a^n = a^{m-n}$	Use the index law for division to simplify expressions.	Simplify $p^{11} \div p^7$
To simplify (cancel) algebraic fractions, look for common factors in exactly the same way as with numeric fractions.	Simplify an algebraic fraction by cancelling.	Simplify $\dfrac{15n + 19}{3}$
Algebraic fractions are added, subtracted or multiplied using the same rules as fractions involving numbers.	Add and subtract fractions with: • different numerical denominators • simple algebraic denominators. Multiply simple algebraic fractions.	Simplify: **a)** $\dfrac{2a}{3} + \dfrac{a}{7}$ **b)** $\dfrac{2}{c} + \dfrac{6}{c}$ **c)** $\dfrac{2}{x} \times \dfrac{4}{5}$

Algebraic expressions or formulae should be written in their simplest form.	Form an algebraic expression or formula for a given situation.	Find an expression for the shaded area:

6 Transformations

You will learn how to:

- Transform points and 2D shapes by combinations of reflections, translations and rotations.
- Identify and describe a transformation (reflections, translations, rotations and combinations of these) given an object and its image.
- Recognise and explain that after any combination of reflections, translations and rotations the image is congruent to the object.
- Enlarge 2D shapes, from a centre of enlargement (outside, on or inside the shape) with a positive integer scale factor. Identify an enlargement, centre of enlargement and scale factor.
- Analyse and describe changes in perimeter and area of squares and rectangles when side lengths are enlarged by a positive integer scale factor.

Starting point

Do you remember...

- how to translate a shape with a vector?

  For example, translate shape A with vector $\begin{pmatrix} -6 \\ -4 \end{pmatrix}$.

- how to reflect a shape in a line?

  For example, reflect shape A in the line $y = 4$.

- how to rotate a shape about a point?

  For example, rotate shape A about the point $(8, 9)$ by 90° in an anticlockwise direction.

- how to enlarge a shape using a given centre?

  For example, enlarge shape A with scale factor 2 and centre $(8, 9)$

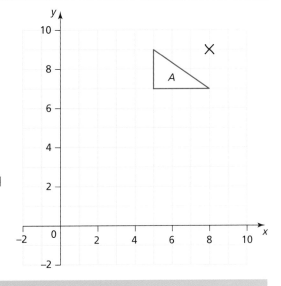

This will also be helpful when...

- you enlarge shapes with a fractional or negative scale factor.

6.0 Getting started

A single move by a knight in a game of chess is a translation of any one of:

$A\begin{pmatrix} 1 \\ 2 \end{pmatrix}$, $B\begin{pmatrix} 1 \\ -2 \end{pmatrix}$, $C\begin{pmatrix} 2 \\ 1 \end{pmatrix}$, $D\begin{pmatrix} 2 \\ -1 \end{pmatrix}$, $E\begin{pmatrix} -1 \\ 2 \end{pmatrix}$, $F\begin{pmatrix} -1 \\ -2 \end{pmatrix}$, $G\begin{pmatrix} -2 \\ 1 \end{pmatrix}$ or $H\begin{pmatrix} -2 \\ -1 \end{pmatrix}$

The diagram shows a chessboard.

a) Find one possible route a knight can take from the square marked X to the square marked Y.

Give your answer in terms of the vectors A, B, C, D, E, F, G and H.

b) Find the shortest route a knight can take from Y to X.

6.1 Describing transformations

Key terms

To describe a transformation mathematically you need to be very precise.

- Use a **vector** to describe a translation.
- To describe a reflection, give the equation of the mirror line.
- To describe a rotation, give the coordinates of the **centre of rotation**, the angle through which the object has been rotated and the direction of the rotation (clockwise or anticlockwise).

Worked example 1

The diagram shows four flags.

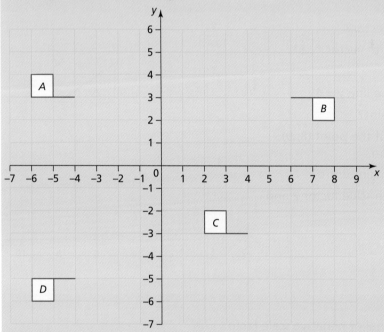

Describe the transformation taking:

a) flag *A* to flag *B*

b) flag *A* to flag *C*

c) flag *A* to flag *D*.

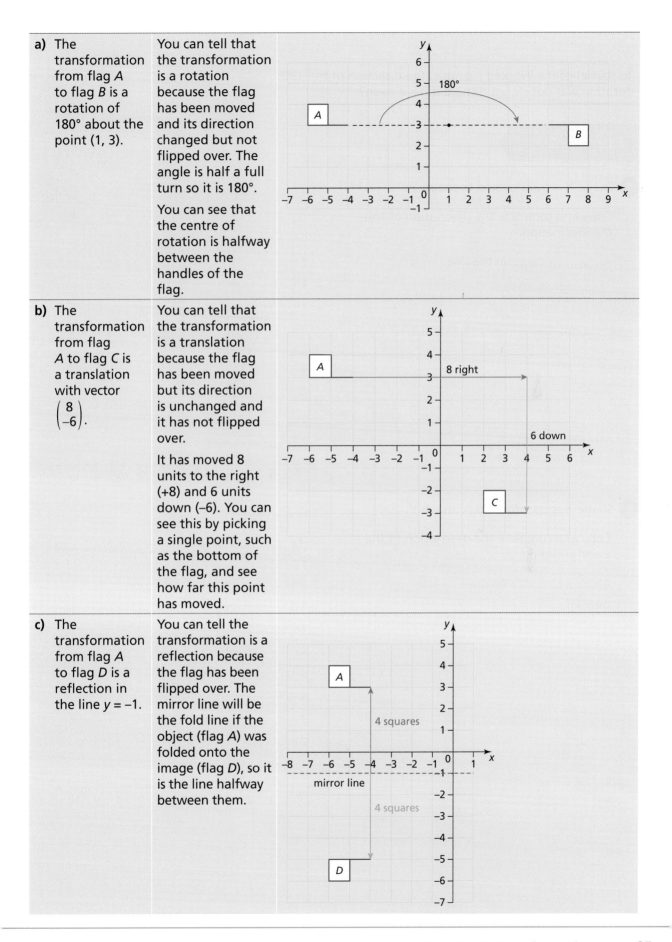

a) The transformation from flag *A* to flag *B* is a rotation of 180° about the point (1, 3).

You can tell that the transformation is a rotation because the flag has been moved and its direction changed but not flipped over. The angle is half a full turn so it is 180°.

You can see that the centre of rotation is halfway between the handles of the flag.

b) The transformation from flag *A* to flag *C* is a translation with vector $\begin{pmatrix} 8 \\ -6 \end{pmatrix}$.

You can tell that the transformation is a translation because the flag has been moved but its direction is unchanged and it has not flipped over.

It has moved 8 units to the right (+8) and 6 units down (–6). You can see this by picking a single point, such as the bottom of the flag, and see how far this point has moved.

c) The transformation from flag *A* to flag *D* is a reflection in the line *y* = –1.

You can tell the transformation is a reflection because the flag has been flipped over. The mirror line will be the fold line if the object (flag *A*) was folded onto the image (flag *D*), so it is the line halfway between them.

In part a) of the Worked example, the direction of the 180°
rotation was not stated. Should it have been?

Exercise 1

1 Shape *X* is transformed to shape *Y*.

Copy and complete the description of this
transformation.

A .... in the line ....

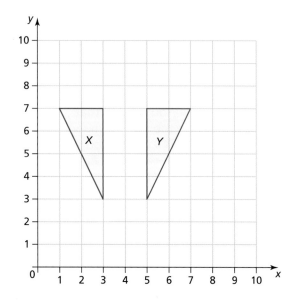

2 Shape *X* is transformed to shape *Y*.

Copy and complete the description of the
transformation.

A .... centre ....
through ....

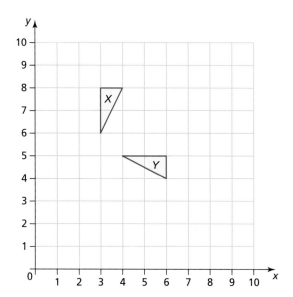

3 Shape *X* is transformed to shape *Y*.

Copy and complete the description of the transformation.

A .... with vector

....

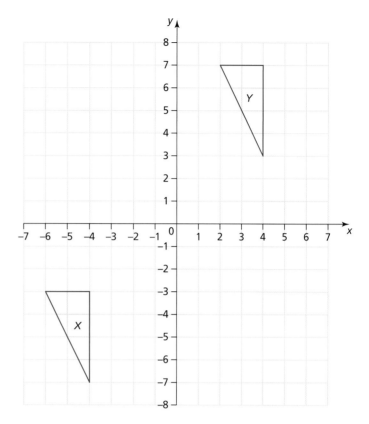

4 In each part of this question, shape *X* is transformed to shape *Y*. Describe each transformation mathematically.

a)

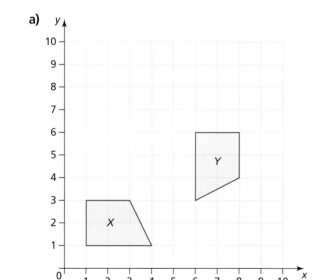

b)

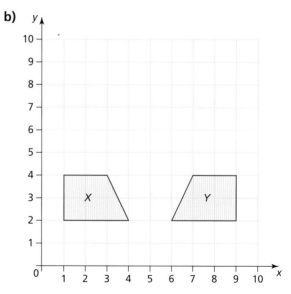

c)

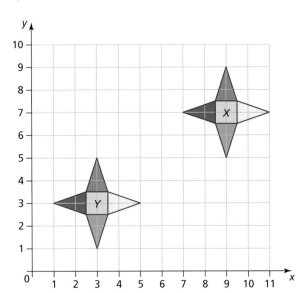

d)

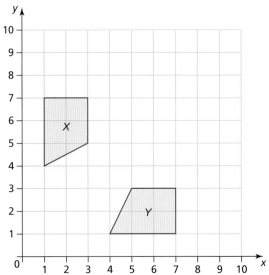

5 The diagram shows four shapes *A*, *B*, *C* and *D*.

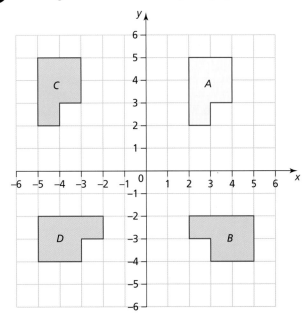

Describe the single transformation that maps:

a) shape *A* to shape *B*   **b)** shape *A* to shape *C*   **c)** shape *A* to shape *D*.

6 Some of the labelled shapes below are transformations of the red trapezium.

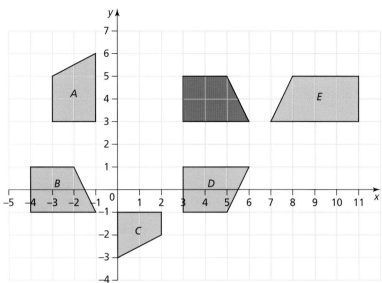

a) For each shape, *A* to *E*, state if it is congruent or not congruent to the red trapezium.

b) For each shape that is congruent to the red trapezium, describe mathematically the transformation from the red trapezium.

7 The diagram shows six shapes.

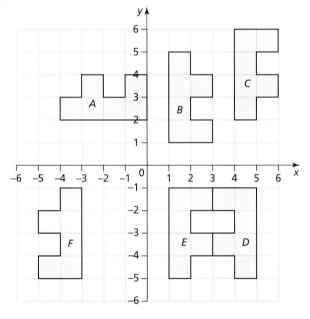

Describe fully the single transformation that maps:

a) shape *A* to shape *B*

b) shape *B* to shape *E*

c) shape *C* to shape *E*

d) shape *E* to shape *F*

e) shape *E* to shape *D*

f) shape *B* to shape *D*.

8 The transformation which takes shape *P* to shape *Q* is a rotation of 90° anticlockwise about point (3,6). Describe the transformation which takes shape *Q* to shape *P*.

9 The diagram shows a square, with corners at (2, 4), (4, 4), (4, 6) and (2, 6), and it's image after a transformation.

Tomas says the transformation is a reflection in the line *y = x*.

a) Explain why Tomas might not be correct.

b) What could be added to the diagram to help determine exactly which transformation has taken place?

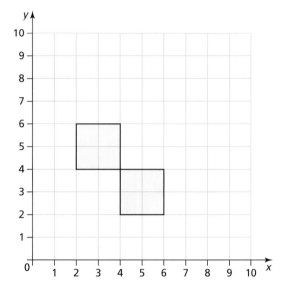

10 Two shapes, *P* and *Q*, are drawn on a grid.

a) A single transformation maps *P* to *Q*. Explain why this transformation cannot be a rotation.

b) Describe two possible transformations that could map *P* to *Q*.

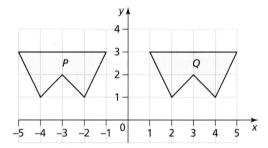

▼▼▼ Thinking and working mathematically activity

Look at the octagon drawn on a grid.

• Describe all the transformations that will transform the octagon onto itself.

• How can you be convinced that you have them all?

• Does a rectangle have as many transformations that maps itself to itself? Give a reason for your answer.

• Describe a shape which has more transformations that map itself to itself than the octagon.

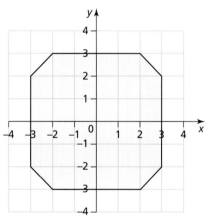

6.2 Combining transformations

Key terms

....

Transformations can be **combined** by carrying out two or more transformations, one after another.

Worked example 2

The diagram shows a grid and a triangle X.

The image of triangle X when it is reflected in the y-axis is triangle Y.

The image of triangle Y when it is reflected in the x-axis is triangle Z.

a) Draw the position of triangle Z on the grid.

b) Which single transformation takes triangle X to triangle Z?

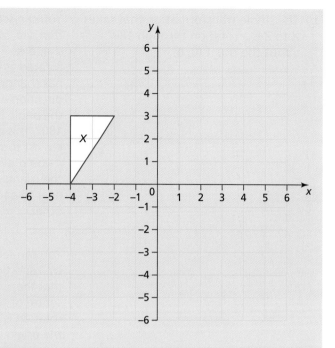

a)

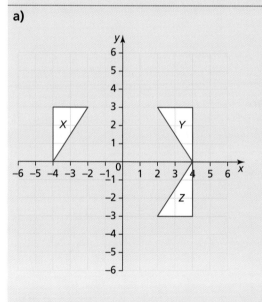

First reflect in the y-axis. Make sure that each point is the same distance from the mirror line as its reflection.

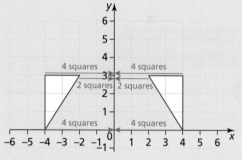

Then reflect this image in the x-axis. Again make sure that each point is the same distance from the mirror line as its reflection.

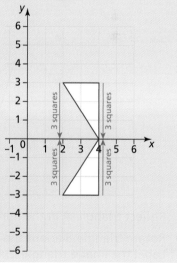

b) The single transformation that takes X to Z is a rotation through 180° about the point (0, 0).

You can see that the shape is rotated as it has changed direction and the angle is 180°. This is easiest to see if you draw the triangles near to each other.

You can see by inspection that the centre of rotation is at the origin.

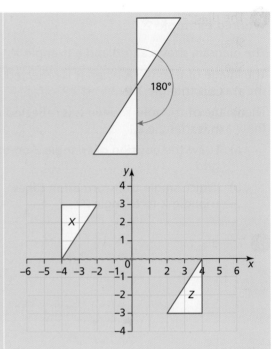

Exercise 2

1 The diagram shows a triangle A drawn on squared paper.

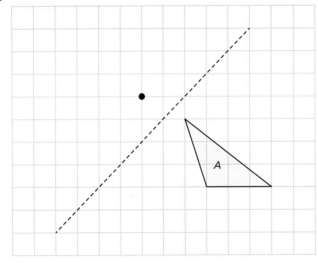

a) Reflect triangle A in the mirror line. Label the image triangle B.

b) Rotate triangle B about the point shown by 90° in a clockwise direction. Label the image triangle C.

> **Think about**
>
> If you change the order of the transformations, would triangle C be in the same place?

2 The diagram shows a shape C drawn on a grid.

Shape C is reflected in the x-axis to give shape D.

Shape D is reflected in the y-axis to give shape E.

a) Copy the diagram. Draw the position of shape E.

b) Describe fully a single transformation that maps shape C to shape E.

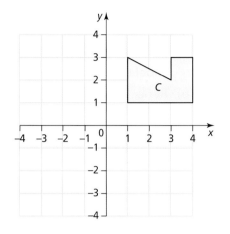

3 The diagram shows a shape S on a grid.

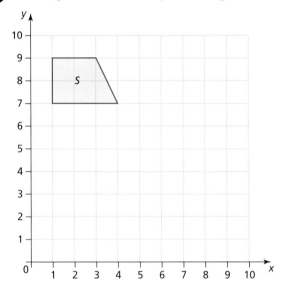

Copy the diagram.

a) Reflect S in the line y = 5. Label the image T.

b) Translate T with vector $\begin{pmatrix} 5 \\ 4 \end{pmatrix}$. Label the image U.

c) Is U congruent with S?

4 Draw another copy of the diagram in question 3.

Shape S is rotated by 90° clockwise about (4, 7). The image is X.

Shape X is then reflected in the line y = 7. The image is Y.

a) Will Y be congruent to S? Give a reason for your answer.

b) Show the position of Y on your grid.

5 Draw another copy of the diagram in question 3.

a) Enlarge S with scale factor 2 from the point (1, 9). Label the image G.

b) Reflect G in the line y = 6. Label the image H.

c) Is S congruent with H?

6 The diagram shows a shape X on a grid. Copy the diagram.

a) Reflect X in the line y = 4. Label the image Y.

b) Reflect Y in the line y = −1. Label the image Z.

c) Describe mathematically the single transformation which takes X to Z.

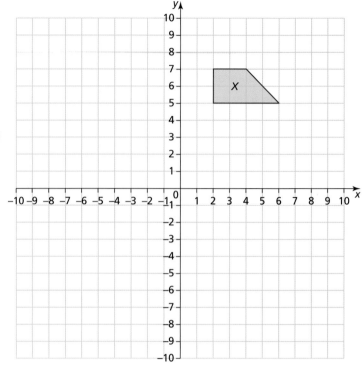

7 Make another copy of the diagram in question 6.

a) Translate X with vector $\begin{pmatrix} 4 \\ 0 \end{pmatrix}$. Label the image P.

b) Reflect P in the line x = 4. Label the image Q.

c) Describe mathematically the single transformation which takes X to Q.

8 Make another copy of the diagram in question 6.

a) Rotate X through 90° clockwise about the point (2, 3). Label the image M.

b) Translate M with vector $\begin{pmatrix} -2 \\ 2 \end{pmatrix}$. Label the image N.

c) Describe mathematically the single transformation which takes X to N.

9 Here is a pentagon A that 'points' to the right.

A is translated by a vector to give B.

B is then reflected in a vertical line to give C.

Decide if each of these statements is true or false.

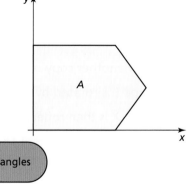

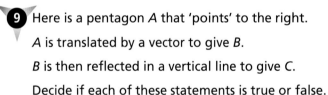

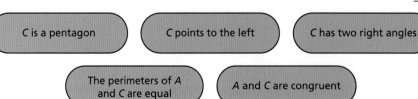

C is a pentagon

C points to the left

C has two right angles

The perimeters of A and C are equal

A and C are congruent

Thinking and working mathematically activity

For this activity you will need a coordinate grid with axes labelled from –6 to 6.

- Investigate what transformation is equivalent to a reflection in a vertical line followed by a reflection in a horizontal line. Does the order matter?

- Investigate what transformation is equivalent to two translations.

- Investigate what transformation is equivalent to a reflection in two parallel vertical (or horizontal) lines.

10 A shape *F* is translated with vector $\begin{pmatrix} -3 \\ -5 \end{pmatrix}$. The image is labelled *G*.

G is translated with vector $\begin{pmatrix} 8 \\ 10 \end{pmatrix}$. The image is labelled *H*.

Describe mathematically the single transformation which takes *F* to *G*.

11 In a video game, a shape is moved using a combination of rotations and reflections. The rotations can only be centred on a vertex of the shape.

Write down a combination of transformations that would result in shape *A* moving to shape *A'*.

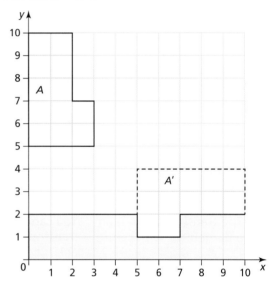

6.3 Enlargements

Key terms

To describe an enlargement, give the **scale factor** and the coordinates of the **centre of enlargement**.

Thinking and working mathematically activity

- Draw a pair of axes on a coordinate grid and label them from –2 to 16.

- Draw a rectangle *R* with vertices *A*(1, 1), *B*(3, 1), *C*(3, 2) and *D*(1, 2).

- Enlarge *R* using different whole number scale factors.

- Investigate how the perimeter of rectangle *R* changes when it is enlarged.

- Investigate how the area of rectangle *R* changes when it is enlarged.

- Test your rules by considering enlargements of a different shape.

Worked example 3

The diagram shows a transformation of flag P to flag Q.

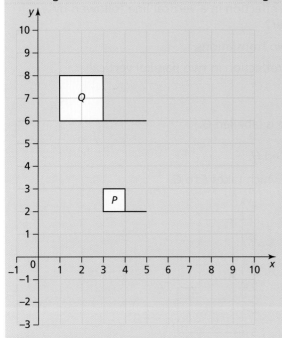

a) Describe fully the transformation of flag P to flag Q.

b) Ramatu says flag Q is not congruent to flag P. Is Ramatu correct? Explain your answer.

a) The transformation is an enlargement with centre $(5, -2)$ and scale factor 2.	You can tell the transformation is an enlargement because the size of the flag has changed.	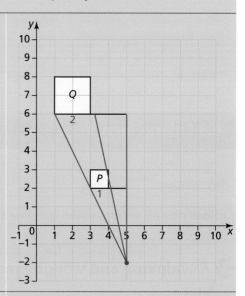
	To find the centre of the enlargement join corresponding points and extend the lines until they meet at the point $(5, -2)$.	
	The side of the original flag is 1 square and the side of the enlarged flag it is 2 squares, so the scale factor of the enlargement is 2.	
b) Ramatu is correct. The two flags are of different sizes so they are not congruent.	Shapes are congruent if they have exactly the same shape and size.	

Worked example 4

A triangle has perimeter 12 cm and area 6 cm².

The triangle is enlarged with scale factor 3.

Find:

a) the perimeter of the image   **b)** the area of the image.

a) Perimeter = 12 × 3 = 36 cm	When a shape is enlarged by a scale factor k, the perimeter is multiplied by k.	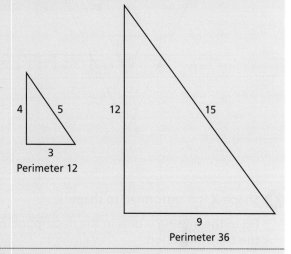
b) Area = 6 × 3² = 6 × 9 = 54 cm²	When a shape is enlarged by a scale factor k, the area is multiplied by k^2.	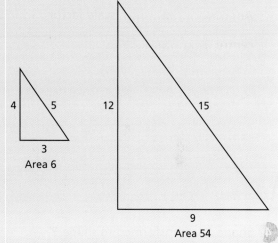

1 A quadrilateral Q is shown on the grid.

  a) Copy the diagram and draw the enlargement of Q with scale factor 3, centre (0, 0). Label the enlargement R.

  b) Draw the enlargement of Q with scale factor 2, centre (0, 2). Label the enlargement S.

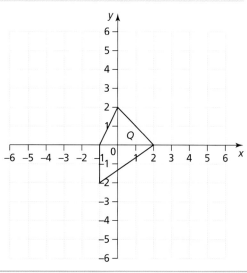

2 In each diagram *N* is an enlargement of *M*. Find the scale factor and centre of enlargement in each case.

a)

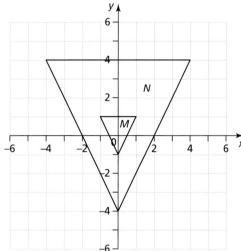

b)

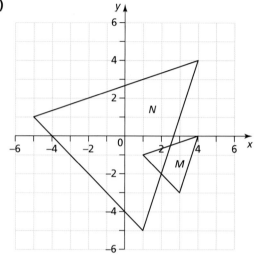

3 Shape *X* is transformed to shape *Y*.

Fill in the missing words in the description of the transformation.

An .... scale factor ....

centre ....

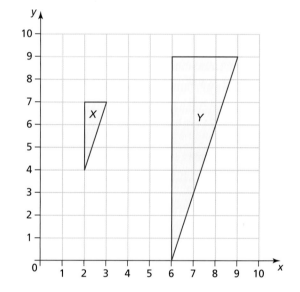

4 Shape *X* is transformed to shape *Y*.

Describe the single transformation mathematically.

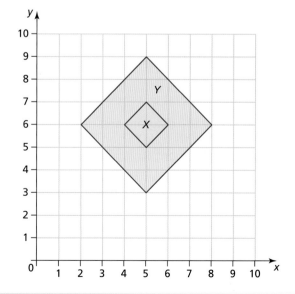

5 The diagram shows shape *A* transformed to shape *B*.

Describe mathematically the single transformation that changes shape *A* to shape *B*.

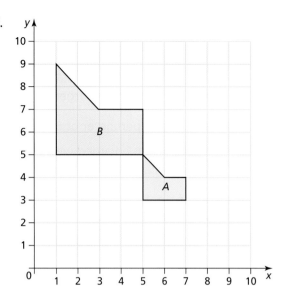

6 The diagram shows two arrows, *P* and *Q*.

Describe the single transformation that maps *P* to *Q*.

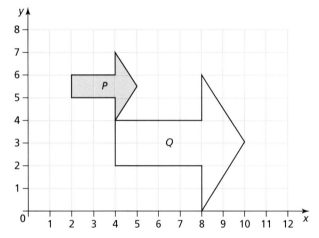

7 The diagram shows quadrilaterals *P*, *Q* and *R*.

Describe the single transformation that maps:

a) *P* to *Q*   b) *P* to *R*

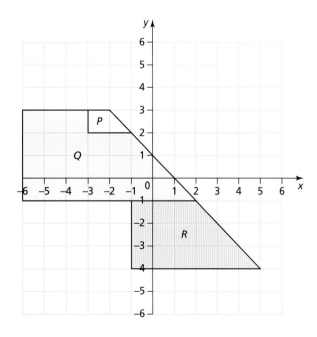

8 The perimeter of a trapezium is 17 cm. The area of the trapezium is 25 cm².

The trapezium is enlarged with a scale factor of 3.

   a) Find the perimeter of the enlarged shape.

   b) Find the area of the enlarged shape.

9 A rectangle has a perimeter of 8 cm and an area of 3 cm².

   a) The rectangle is enlarged with a scale factor of 4. Find the perimeter of the enlarged rectangle.

   b) The rectangle is enlarged with a scale factor of 5. Find the area of the enlarged rectangle.

10 The area of a polygon is 15 cm². The polygon is enlarged by a scale factor of 2. Find the area of the enlarged polygon.

11 Ellie draws a shape and enlarges it. The original is 3 cm wide. The enlargement is 12 cm wide.

The area of the original shape is 20 cm².

Ellie says the area of the enlarged shape will be 80 cm².

Is Ellie correct? Give a reason for your answer.

12 Max draws a rectangle, *T*, with a length of 3 cm.

Tina draws a rectangle, *L*, with a length of 6 cm.

Max says, 'Our rectangles are in the ratio of 1 : 2.'

Tina says, 'They are in the ratio of 1 : 4.'

Explain how they can both be correct.

Consolidation exercise

1 The diagram shows some triangles drawn on a grid.

Describe fully each single transformation from the shaded triangle to:

   a) shape *A*   b) shape *B*   c) shape *C*

   d) shape *D*   e) shape *E*   f) shape *F*.

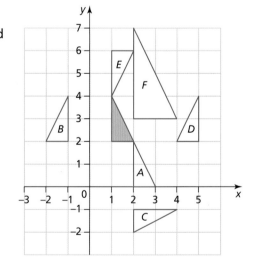

2 Copy the diagram.

a) Reflect *T* in the *y*-axis.
    Label the image *A*.

b) Rotate *A* by 90° clockwise about the point
    (−5, −1). Label the image *B*.

c) Translate *B* with the vector $\begin{pmatrix} 3 \\ 1 \end{pmatrix}$.
    Label the image *C*.

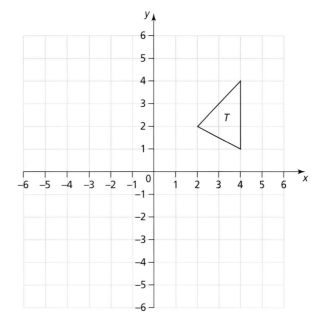

3 The diagram shows a grid with a
trapezium *X*. Make two copies of the
diagram.

a) On the first copy of the diagram,
    reflect *X* in the line *x* = −5 and
    then reflect this image in the
    y-axis. Label the final image *Y*.

b) Which single transformation
    takes *X* to *Y*?

c) On the second copy of the
    diagram, reflect *X* in the line
    x = −5 and then reflect this
    image in the line *y* = 5.
    Label the final image *Z*.

d) Which single transformation
    takes *X* to *Z*?

e) Explain why a combination of
    two reflections can be the same
    as a translation or a rotation but
    not another reflection.

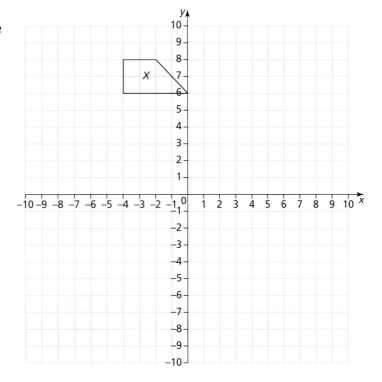

The diagram shows three shapes A, B and C.

a) Describe fully the single transformation that takes shape A to shape B.

b) Describe fully the single transformation that takes shape A to shape C.

c) Clive said all three shapes are congruent. Is Clive correct? Explain your answer.

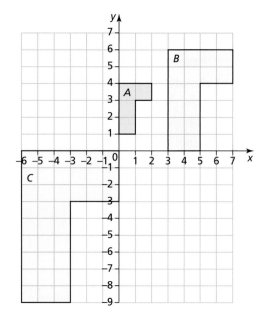

The rectangle Q is an enlargement of the rectangle P with scale factor 3.
The rectangle Q is 12 cm long and 9 cm high. Find the perimeter of rectangle P.

6 Andrew draws a polygon with a perimeter of 12 cm and an area of 5 cm².

Sophia draws an enlargement of Andrew's polygon that has a perimeter of 48 cm.

What is the area of Sophia's polygon?

7 A forklift truck driver has to move an isosceles triangle shaped wooden crate from X to Y.

The truck can only move using translations and rotations.

Using translations and rotations, describe how the driver can move the crate from X to Y without passing through the blue walls.

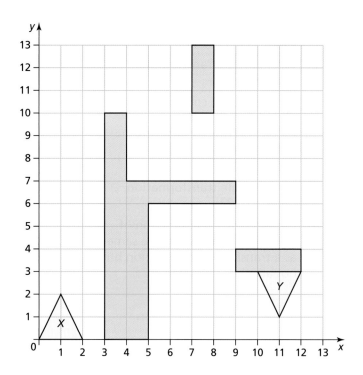

End of chapter reflection

You should know that...	You should be able to...	Such as...
Transformations need to be defined precisely.	Describe mathematically: reflections, rotations, translations and enlargements.	Define the transformation which takes triangle *A* to triangle *B*.
Transformations can be combined.	Combine transformations and describe the single transformation which produces the same image.	Triangle *P* is rotated through 90° clockwise about (0, 0). This image is then reflected in the *y*-axis to the final image *Q*. What single transformation would take *P* to *Q*?
When a shape is enlarged with a scale factor k: • the perimeter of the shape is multiplied by k • the area of the shape is multiplied by k^2.	Calculate the lengths and areas in an enlarged shape if you know the corresponding lengths and area of the original shape.	A shape has length 5 cm and area 8 cm^2. The shape is enlarged with scale factor 3. **a)** Find the length of the enlarged shape. **b)** Find the area of the enlarged shape.

7

Presenting and interpreting data 1

You will learn how to:

- Record, organise and represent categorical, discrete and continuous data. Choose and explain which representation to use in a given situation:
  o frequency polygons
  o stem-and-leaf and back-to-back stem-and-leaf diagrams
- Interpret data, identifying patterns, trends and relationships, within and between data sets, to answer statistical questions. Make informal inferences and generalisations, identifying wrong or misleading information.

Starting point

Do you remember...

- how to use inequality signs to define intervals when grouping continuous data?

  For example, $3 < n \leq 9$. What is the largest value n could be? What is the smallest value n could be?

- how to draw and interpret frequency diagrams for grouped discrete and continuous data?

  For example, draw a frequency diagram to represent the data in the table below.
  Write down the modal class of the data.

Height, h (m)	Frequency
$0 < h \leq 4$	5
$4 < h \leq 8$	10
$8 < h \leq 12$	8
$12 < h \leq 16$	3

- how to draw and interpret simple line graphs? For example, what is the temperature of the liquid after 15 minutes?

- how to draw and interpret simple stem-and-leaf diagrams? For example, draw a stem-and-leaf diagram to show these numbers:

23	25	27	30	34	38	39
41	42	46	46	51	55	63

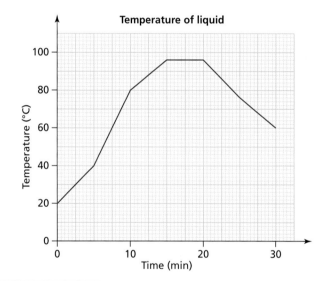

This will also be helpful when...

- you learn how to find the mean from a grouped frequency table
- you use the mode and range to compare two distributions
- you draw statistical conclusions about conjectures.

7.0 Getting started

The number of pets owned by two different classes of students are shown in the frequency tables below.

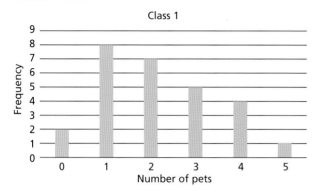

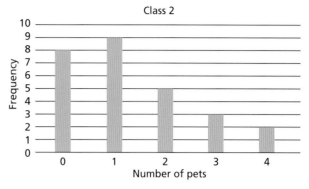

Which class owned the most pets?

What was the modal number of pets owned by each class?

How could you find the median number of pets?

Compare the two frequency diagrams. What can you say about the number of pets owned by the classes?

7.1 Frequency polygons

Key terms

A **frequency polygon** is a graph formed by drawing straight lines between points representing the frequency of the class intervals plotted at their **midpoints**.

A frequency diagram (bars) and a frequency polygon (blue lines) representing the length in centimetres of a sample of 20 flowers are both drawn on the same axis.

To calculate the **midpoint** of a class interval, add the upper and lower boundary of the interval and divide by 2.

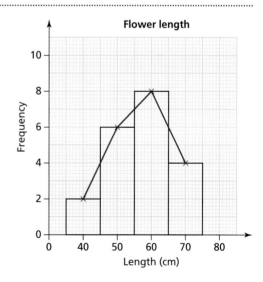

Frequency polygons can be **open** or **closed**. A closed frequency polygon is when the points are drawn down to the horizontal axis at each end.

Open frequency polygon

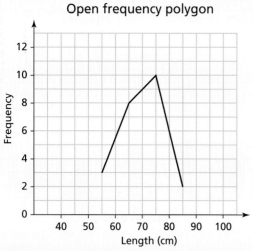

Closed frequency polygon

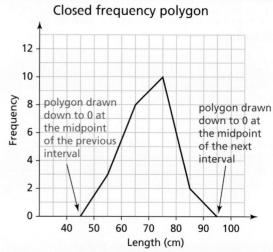

Frequency polygons are useful as they allow two sets of data to be compared on the same graph.

Worked example 1

A gardener measures the diameters of 50 apples from his orchard.
The results are shown in the frequency table.

Diameter, d (mm)	Frequency
$50 \leq d < 54$	4
$54 \leq d < 58$	10
$58 \leq d < 62$	18
$62 \leq d < 66$	12
$66 \leq d < 70$	6

a) Write down the modal class.

b) Draw a frequency polygon to show the information.

---

a) $58 \leq d < 62$

The modal class is the class interval with the highest frequency.

The highest frequency is 18 and this corresponds to the class interval $58 \leq d < 62$.

b)

Diameter, d (mm)	Frequency	Midpoint
$50 \leq d < 54$	4	$\dfrac{50 + 54}{2} = 52$
$54 \leq d < 58$	10	$\dfrac{54 + 58}{2} = 56$
$58 \leq d < 62$	18	$\dfrac{58 + 62}{2} = 60$
$62 \leq d < 66$	12	$\dfrac{62 + 66}{2} = 64$
$66 \leq d < 70$	6	$\dfrac{66 + 70}{2} = 68$

First, find the midpoint of each class interval.

Add the upper and lower bounds of each interval and divide by 2.

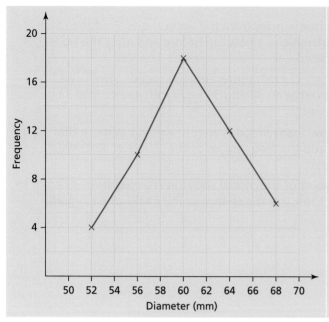

Plot the calculated midpoints against the frequency.

The data are continuous so the frequency polygon should have a continuous scale.

Remember to label the axes.

Frequency is always plotted on the vertical axis.

Draw straight lines between each point to complete the frequency polygon.

Note this is an open frequency polygon.

Worked example 2

A council places speed awareness signs on a busy road outside a school. The council measures the speed of a sample of vehicles before and after the signs are installed to compare the impact of the signs. The results are shown in the two frequency polygons.

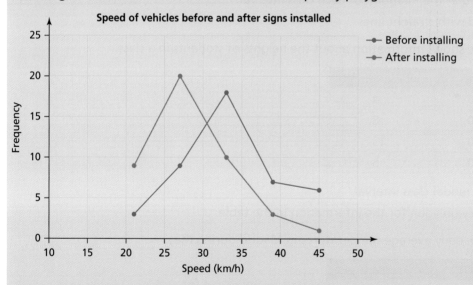

a) Find how many vehicles had their speeds measured in each sample.

b) Find how many vehicles were travelling at over 36 km/h before the speed awareness signs were installed.

c) Jane says that the speed awareness signs have been successful in reducing the speed of vehicles passing the school. Is Jane correct? Give a reason for you answer.

a) Before: $3 + 9 + 18 + 7 + 6 = 43$ After: $9 + 20 + 10 + 3 + 1 = 43$	Read the frequencies from the vertical axis. Find the total number of vehicles measured before and after the signs were installed.
b) $7 + 6 = 13$ vehicles	The point at 33 km/h represents the speeds of vehicles travelling between 30 and 36 km/h (remember that 33 is the midpoint). To find how many vehicles were travelling at more than 36 km/h, add the frequencies of the points at 39 km/h and 45 km/h.
c) Jane is correct. The modal speed has reduced from 33 km/h to 27 km/h.	The midpoint of the modal class has reduced from 33 km/h to 27 km/h. These are the speeds at the peaks of each graph.

Exercise 1

1 Write down whether each statement about frequency polygons is true or false.

    a) Frequency is plotted on the horizontal axis.

    b) Points are plotted at the start of each class interval.

    c) Points are joined with a smooth curve.

    d) Points are plotted at the midpoint of each class interval.

    e) Points are joined with straight lines.

2 The frequency table shows information about the heights of students in a class.

Height, h (cm)	Frequency
$140 \leq h < 150$	4
$150 \leq h < 160$	8
$160 \leq h < 170$	15
$170 \leq h < 180$	5

    a) Write down the modal class interval.

    b) Draw a frequency polygon for the information in the table.

3 The table shows the daily average temperatures in Madrid during May.

Temperature, t (°C)	Frequency
$10 \leq t < 12$	1
$12 \leq t < 14$	
	12
$16 \leq t < 18$	8
$18 \leq t < 20$	4

    a) Write down the missing class interval.

    b) Write down the missing frequency.

    c) Draw a frequency polygon to show the data.

4 The table shows the times taken for 40 boys and 40 girls to complete a quiz.

Time, t (minutes)	Frequency boys	Frequency girls
$20 \leq t < 24$	4	6
$24 \leq t < 28$	7	9
$28 \leq t < 32$	9	14
$32 \leq t < 36$	17	9
$38 \leq t < 42$	3	2

a) Draw two frequency polygons on the same graph to show this data.

b) Compare the times taken for the boys and girls.

5 The table shows the times taken for a sample of students to travel to school.

Time, t (minutes)	Frequency
$5 < t \leq 10$	24
$10 < t \leq 15$	38
$15 < t \leq 20$	57
$20 < t \leq 25$	12
$25 < t \leq 30$	9

a) Draw a frequency polygon to show this information.

b) Find how many students took longer than 20 minutes to travel to school.

c) A student is selected at random. Find the probability that they take longer than 25 minutes to travel to school. Give your answer as a fraction in its simplest form.

6 The frequency polygon shows the length of a sample of fish caught in a lake during a competition.

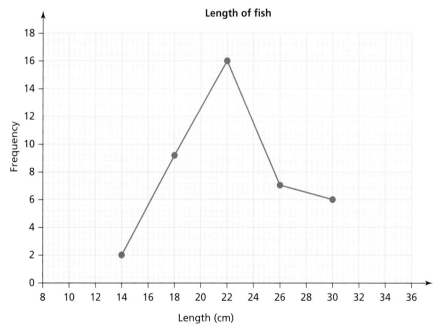

a) Copy and complete the table. The first row has been done for you.

Length, l (cm)	Midpoint	Frequency
$12 \leq l < 16$	14	2

b) Find how many fish were in the sample.

c) Write down the modal class interval.

d) Find the percentage of the sample that were less than 20 cm in length.

e) Explain why we cannot say how many fish were exactly 22 cm in length.

7 Real data question The frequency polygons show the marathon finishing times for men and women at the Rio 2016 Olympic Games.

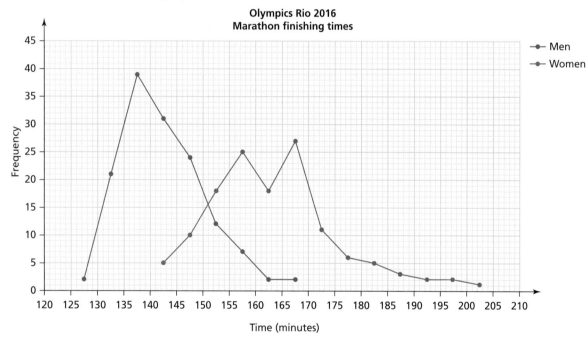

Source: International Olympic Committee

a) Write down the modal class intervals for men and women.

b) Find how many male athletes finished the race.

c) Find the percentage of female athletes who were faster than 2 hours and 25 minutes.

d) Phyllis says that more than 44% of men finished in under 2 hours and 20 minutes. Is Phyllis correct? Explain your answer.

Thinking and working mathematically activity

Cynthia travels to work by bus one month and by car the following month. She notes how many minutes late she was on each occasion. The table shows the results.

Time, t (minutes)	Bus	Car
$0 \leq t < 4$	6	2
$4 \leq t < 8$	12	8
$8 \leq t < 12$	8	15
$12 \leq t < 16$	4	4
$16 \leq t < 20$	0	2

Draw two frequency polygons on the same graph to show this information.

Make some conclusions about the differences Cynthia discovered between travelling by bus and car.

Did you know?

Florence Nightingale became the first female member of the Royal Statistical Society in 1858. She once wrote, 'Statistics is the most important science in the whole world.'

7.2 Back-to-back stem-and-leaf diagrams

Key terms

A **back-to-back stem-and-leaf diagram** is a type of stem-and-leaf diagram that can be used to compare two sets of data.

Here is a back-to-back stem-and-leaf diagram showing the time that 11-year-olds and 14-year-olds took to run the same race.

Notice that the data for 11-year-olds shares a common stem with the data for 14-year-olds.

The leaves are written in order of size. The smallest values in each row are always written nearest the stem.

Race times for 11-year-olds and 14-year-olds

11-year-olds		14-year-olds
9 8	8	2 5 6 9
6 5 2 1	9	0 4 5 6 7 7
9 8 8 5 2 0	10	1 2 3 5 8
7 7 7 6	11	3 7
9 5 4	12	
5 1	13	8

Key 9 | 8 = 89 seconds   **Key** 8 | 2 = 82 seconds

Think about

How do the race times for 11-year-olds compare with those for 14-year-olds?

Worked example 3

Some children and some adults tried to guess the number of buttons in a jar.

The guesses for the children are shown in the back-to-back stem-and-leaf diagram.

Adults		Children
	14	0 5 6
	15	0 0 0 5 7
	16	0 2 5 5 8
	17	5 5 8
	18	0 1 5 5 6
	19	0 2 5
	20	0 0 0 5

Key   Key 14 | 0 = 140 buttons

The guesses for the adults are:

179	146	200	150	165	152	155	179	149	195	177
145	154	166	192	184	180	148	162	153	160	

a) Complete the back-to-back stem-and-leaf diagram by including the guesses for the adults.

b) Find the median number of guesses for the adults and the children.

c) The actual number of buttons in the jar was 171. Find who got the closest to the correct number.

a)

Adults		Children
9 8 6 5	**14**	0 5 6
5 4 3 2 0	**15**	0 0 0 5 7
6 5 2 0	**16**	0 2 5 5 8
9 9 7	**17**	5 5 8
4 0	**18**	0 1 5 5 6
5 2	**19**	0 2 5
0	**20**	0 0 0 5

Key 5 | 14 = 145 buttons   Key 14 | 0 = 140 buttons

Start by writing the data values in order of size.

Enter each number onto the diagram, remembering that the smallest values in each row are written closest to the stem.

Remember to complete the key.

Ordered list of data values:

145   146   148   149
150   152   153   154 155
160   162   165   166
177   179   179
180   184
192   195
200

b) 21 adults had a guess.

The median is in position $\frac{21+1}{2}$ = 11

So the median for adults is 162

28 children had a guess.

The median is in position $\frac{28+1}{2}$ = 14.5, that is halfway between the 14th and 15th values.

These are both 175, so the median for children is 175.

When there are n values written in order, the median is the value in position $\frac{n+1}{2}$.

	Adults	
9 8 6 5	**14**	
5 4 3 2 0	**15**	
6 5 ②0	**16**	
9 9 7	**17**	
4 0	**18**	
5 2	**19**	
0	**20**	

Key 5 | 14 = 145 buttons

	Children
14	0 5 6
15	0 0 0 5 7
16	0 2 5 5 8
17	⑤⑤8
18	0 1 5 5 6
19	0 2 5
20	0 0 0 5

Key 14 | 0 = 140 buttons

c) The closest guess to 171 was made by the child who guessed 168.	Pick out the numbers for children and adults that are closest to 171.	The closest guesses to 171 for adults were 166 and 177: 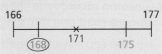 The closest guesses for children were 168 and 175.

Worked example 4

Clare wants to compare how many times men and women have read a newspaper in the past month. She draws this back-to-back stem-and-leaf diagram to show the number of newspapers read by a sample of 15 men and 15 women.

```
    Women                           Men
          9 6 4 3 0  │ 0 │  0 0 1 2 7
        8 7 5 3 1 0  │ 1 │  3 6
              7 6 4 2  │ 2 │  0 0 1 6
                       │ 3 │  0 0 1 1
```

Key 0 | 1 = 10 newspapers   **Key** 1 | 3 = 13 newspapers

a) Calculate the median and range for the number of newspapers read by men.

The median number of newspapers read by the women is 13.

The range for the number of newspapers read by the women is 27.

b) Compare the number of newspapers read by men and women.

a) The median value is 20 newspapers. The range is: 31 – 0 = 31 newspapers.	There are 15 men. The median value will be the value in the $\frac{15+1}{2}$ = 8th position. The range is the difference between the largest and smallest values.	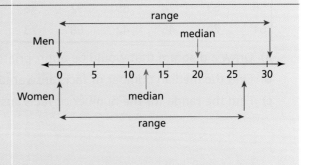

b) The median number of newspapers for men is larger than the median for women. This means that men read more newspapers on average. The range in the number of newspapers is greater for men than for women. This means that there is a greater spread in the number of newspapers read by men.	Compare the values for the median and range. Write your comparisons in the context of the question.	

Exercise 2

1 Giacomo records the speeds of 25 cars passing along a road one morning.

He also records the speeds of another 25 cars passing along the same road in the afternoon.

The back-to-back stem-and-leaf diagram shows his results.

a) How many cars were travelling slower than 40 km/h in the afternoon?

b) Write down the mode of the speeds recorded in the morning.

c) Find how many of the 50 cars were travelling at more than 75 km/h.

d) Find the median of the speeds recorded in the afternoon.

Morning		Afternoon
7 2	3	0 3 8 9
7 7 6 5 4 0	4	2 2 5 7 9
8 6 5 2 1	5	4 5 5 5 8 9
5	6	0 2 4 5 6
6 5 4 2 1 1 1	7	3 4 8
8 5 1 0	8	0 5

Key 2 | 3 = 32 km/h   **Key** 3 | 0 = 30 km/h

2 Here are the times (in seconds) of some boys and girls in a running race.

Boys

52	63	81	66	75	59	77	66	80
64	72	78	58	61	68	72	76	66

Girls

74	79	65	82	87	91	68	77	75	89
86	81	70	93	68	74	80	68	84	

Show this data on a back-to-back stem-and-leaf diagram.

3 A teacher recorded how many multiplication facts two groups of children could answer correctly in four minutes.

Group 1

76	43	65	69	71	84	55	60	82	53
61	54	49	72	67	52	48	59	61	72

Group 2

56	63	74	89	92	48	37	64	62	55
61	70	45	42	68	73	76	82	77	

a) Draw a diagram that would be appropriate for comparing these two sets of data.

b) Find the median number of facts answered correctly by each group.

c) Find the range for the number of facts answered correctly by each group.

End of chapter reflection

You should know that...	You should be able to...	Such as...
Frequency polygons allow two sets of data to be compared on the same graph. To draw a frequency polygon for continuous data, draw straight lines between points representing the frequency of the class intervals plotted at their midpoints	Draw and interpret frequency polygons.	The frequency polygons show the test scores out of 100 of two groups of students. 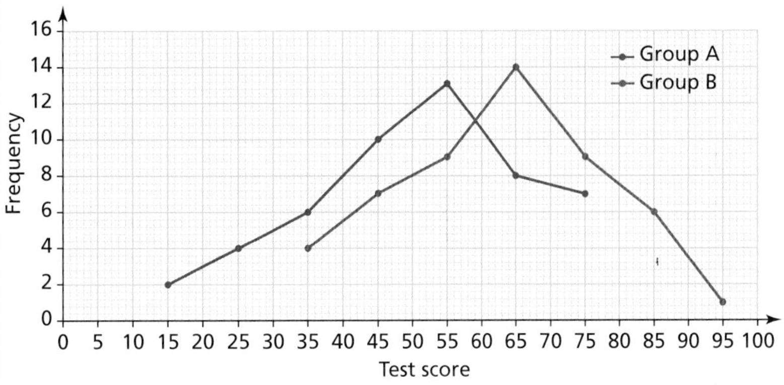 **a)** Find how many students were in group A. **b)** Make some comparisons between the two graphs.
A back-to-back stem-and-leaf diagram shows two sets of data on a common stem.	Draw and interpret a back-to-back stem-and-leaf diagram.	The diagram shows the ages of men and women at a concert. Women | | Men 8 6 5 3 2 | **4** | 3 8 5 4 1 0 | **5** | 0 2 2 6 7 8 5 4 3 2 2 | **6** | 1 2 5 6 7 9 9 7 6 2 0 | **7** | 0 3 4 5 8 7 6 6 5 2 1 | **8** | 1 4 6 7 8 8 9 3 1 | **9** | 0 4 5 **Key** 2 | **4** = 42 years **Key** **4** | 3 = 43 years Find how many people at the concert were older than 83 years old.

You will learn how to:

- Understand that when a number is rounded there are upper and lower limits for the original number.
- Estimate, multiply and divide decimals by integers and decimals.

Starting point

Do you remember...

- how to round numbers to the nearest 1000, 100, 10 or 1?
  For example, round 576 to the nearest 100.
- how to round numbers to a given number of decimal places?
  For example, round 8.79 to one decimal place.
- how to round numbers to a given number of significant figures?
  For example, round 0.365 to one significant figure.
- how to estimate and multiply decimals by integers and decimals?
  For example, find 0.36×2.4

This will also be helpful when...

- you learn to find upper and lower limits on calculated values.

8.0 Getting started

Many news headlines contain numbers. Look at the examples here.

Do you think these numbers are all exactly correct, or have they been rounded? Suggest why news headlines might use rounded numbers.

For each headline, suggest three possible values of the exact number. Explain your choices.

Find other examples of news headlines containing numbers. Sort them into numbers you think are exact and numbers you think are rounded. Explain your choices.

DAILY NEWS

/ NO. 1234 /

Woman runs 500 km to raise money for charity

30 000 species of animal and plant threatened with extinction

80 countries now using wind to generate electricity

Two million dollars stolen from bank

Key terms

The **limits** (or **bounds**) on a rounded quantity show the possible values of the quantity before it was rounded.

The **lower limit** (or **lower bound**) is the lowest possible value of a quantity before it was rounded.

The **upper limit** (or **upper bound**) is the lowest number that *cannot* be the value of a quantity before it was rounded.

Upper and lower limits can be written using a single inequality. For example, a length is 120 cm to the nearest 10. Then the limits are $115 \leq \text{length} < 125$.

Did you know?

Upper and lower limits are sometimes called 'limits of accuracy'.

Thinking and working mathematically activity

Create a table showing each of the numbers 70 to 90 and each of these numbers after rounding to the nearest ten.

Exact number	Number rounded to the nearest 10
70	70
71	70
72	70
73	70
74	70
75	80

The number of people at a cinema is 80 to the nearest ten. List the possible numbers of people. Write this as an inequality of the form $\square \leq n < \square$, where n is the actual number of people.

Find inequalities of the form $\square \leq n < \square$ to show:

* the actual number of people in a cinema if the rounded value is 230 to the nearest ten
* the actual number of people at a football match if the rounded value is 8700 to the nearest hundred
* the actual number of people living in a town if the rounded value is 15 000 to the nearest thousand.

Explain your reasoning.

Describe a method for working out the possible values of a number that has been rounded to the nearest 1, 10, 100 or 1000.

Worked example 1

a) The number of people at a sports match is 15 000 to the nearest thousand. Write the limits on the number of people, *n*, at the match.

b) The length of a snake is *L* cm. This is 36 cm to the nearest whole number. Write the limits on *L*.

c) In 2009, Usain Bolt set a new world record for the 100 m sprint. He ran 100 m in *t* seconds, where *t* is 9.58 correct to 2 decimal places. Write the limits on *t*.

a) $14\,500 \le n < 15\,500$	The lower limit is the lowest number that rounds up to 15 000 to the nearest 1000. This is 14 500. The upper limit is the lowest number that does *not* round down to 15 000 to the nearest 1000. (It rounds up to 16 000.)	On a number line, mark 15 000 and the numbers that are 1000 above and below. The upper and lower limits are half-way between the marked numbers.
b) $35.5 \le L < 36.5$	35.5 is the lowest number that rounds to 36 to the nearest whole number. 36.5 is the lowest number that does *not* round to 36 to the nearest whole number.	On a number line, mark 36 and the numbers that are 1 above and below. The upper and lower limits are half-way between the marked numbers.
c) $9.575 \le t < 9.585$	9.575 is the lowest number that rounds to 9.58 to 2 decimal places. 9.585 is the lowest number that does *not* round to 9.58 to 2 decimal places. 'To 2 decimal places' means to the nearest 0.01.	On a number line, mark 9.58 and the numbers that are 0.01 above and below. The limits are half-way between the marked numbers.

For a): 14 500 ... 15 500 marked half-way between 14 000, 15 000, 16 000.

For b): 35.5 ... 36.5 marked half-way between 35, 36, 37.

For c): 9.575 ... 9.585 marked half-way between 9.57, 9.58, 9.59.

Worked example 2

The height of a tree is *h* metres. This is 5.6 m after rounding to two significant figures. Write the limits on *h*.

$5.55 \le h < 5.65$	The second significant figure is the tenths figure, so *h* is 5.6 to the nearest tenth. 5.55 is the lowest number that rounds to 5.6 to the nearest tenth. 5.65 is the lowest number that does *not* round to 5.6 to the nearest tenth.	On a number line, mark 5.6 and the numbers 0.1 above and below. The limits are half-way between the marked numbers.

5.55 ... 5.65 marked half-way between 5.5, 5.6, 5.7.

1. Write an inequality showing the limits on *x*, where *x* is:

   a) 70 to the nearest ten
   b) 2300 to the nearest hundred
   c) 6000 to the nearest thousand
   d) 380 to the nearest ten
   e) 1280 to the nearest ten
   f) 55 000 to the nearest thousand
   g) 87 000 to the nearest thousand
   h) 87 000 to the nearest hundred

2. Write an inequality showing the limits on *x*, where *x* is:

   a) 22 to the nearest whole number
   b) 3.8 to one decimal place
   c) 586 to the nearest whole number
   d) 0.92 to two decimal places
   e) 58.2 to one decimal place
   f) 60 to the nearest whole number
   g) 7.31 to two decimal places
   h) 30.0 to one decimal place

3. The number of trees in a forest is 2000, rounded to the nearest hundred. Which of the numbers below are possible numbers of trees in the forest?

   1900   1908   1950   1951   1987   2000

   2007   2026   2050   2051   2995

4. The length of a road, *L* metres, is measured as 3000 m. Write an inequality showing the limits on *L* if is measured to:

   a) the nearest thousand
   b) the nearest hundred
   c) the nearest ten
   d) the nearest whole number

5. The mass, *m* grams, of an ostrich egg is measured. It is 1400 g after rounding to the nearest 50 g.

   Write the limits on *m*.

   > **Tip**
   >
   > If you round to the nearest 50, what are the possible results? Think about the 50s above and below 1400.

6. A number, *p*, is 16.5 after rounding to the nearest 0.5. Which of the inequalities shows the limits on *p*?

   $16 \leq p < 17$ $16.25 \leq p < 16.75$ $16.45 \leq p < 16.55$

7. Decide if each statement is true or false:

   a) Rounding 430 to 1 significant figure is equivalent to rounding it to the nearest hundred.

   b) Rounding 38.2 to 2 significant figures is equivalent to rounding it to the nearest tenth.

   c) Rounding 88.67 to 3 significant figures is equivalent to rounding it to 1 decimal place.

8. Write an inequality showing the limits on *x*, where *x* is:

   a) 800 to 1 significant figure
   b) 800 to 2 significant figures
   c) 0.75 to 2 significant figures
   d) 900 to 3 significant figures

9 Iolany says, 'If a quantity x is 5 after rounding to the nearest whole number, the highest possible value of x is 5.5.'

Helgi says, 'No, the highest possible value is 5.49.'

Khulan says, 'No, the highest possible value is 5.499.'

Explain whether each statement is correct or incorrect.

Think about

What is the highest possible value of x?

10 The number of paperclips, n, in a jar is 40 to the nearest ten.

The total mass of the paperclips is m grams, where m is 40 to the nearest ten.

The limits on the number of paperclips are $35 \leq n < 45$. The limits on the total mass are $35 \leq m < 45$.

The limit on n can also be written as $35 \leq n < 40$, but the limits on m cannot be written as $35 \leq m \leq 44$.

Explain why.

Discuss

Is it possible for a rounded number to equal zero?

8.2 Multiplying and dividing with integers and decimals

Worked example 3

Estimate and then calculate:

a) 0.162×1.9

b) $5.52 \div (-2.4)$

a) $0.162 \times 1.9 \approx 0.2 \times 2 = 0.4$	To estimate, round each number to one significant figure.
$\begin{array}{r} 162 \\ \times\ 19 \\ \hline 1458 \\ 1620 \\ \hline 3078 \end{array}$	To find the exact answer, first ignore the decimal points and find 162×19
$0.162 \times 1.9 = 0.3078$	Use the estimated answer, 0.4, to see where to place the decimal point.
b) $5.28 \div (-2.4) \approx 5 \div (-2) \approx -2$	To estimate, round each number to one significant figure. Ignore any remainder in the division.
$5.28 \div (-2.4) = \dfrac{5.28}{-2.4} = \dfrac{52.8}{-24}$	To find the exact answer, write the division as a fraction. Multiply the numerator and denominator by 10 to change the divisor (-2.4) into an integer.
$\begin{array}{r} 2.\ 2 \\ 24\overline{)52.^48} \end{array}$	Calculate $52.8 \div 24$ using short or long division.
$5.28 \div (-2.4) = -2.2$	$52.8 \div -24$ is the negative of this, so it is -2.2

1 Estimate and then calculate:

   a) 12 × 34   **b)** 46 × 8.2   **c)** −9.6 × 3.4

   d) 36.7 × 7.8   **e)** −2.45 × (−4.8)   **f)** 36.7 × (−0.83)

2 Estimate and then calculate:

   a) 392 ÷ 14   **b)** 37.4 ÷ 11   **c)** −2.04 ÷ (−0.12)

   d) 0.874 ÷ 2.3   **e)** −2.47 ÷ 1.3   **f)** 20.88 ÷ (−5.8)

3 Use the numbers to complete each statement:

   7.9   19   1.9   6.3   12.42

   a) 2.3 × ..... = 4.37   **b)** 3.4 × ..... = 26.86

   c) 11.34 ÷ ..... = 1.8   **d)** ..... ÷ 4.6 = 2.7

> **Tip**
>
> Use estimation to answer this question.

4 Estimate and then calculate:

   a) −8.4 × (−66)   **b)** 23.1 × 0.94   **c)** 23.1 × (−0.78)

   d) 5.7 ÷ 1.5   **e)** −2.66 ÷ (−1.9)   **f)** −2.03 ÷ 0.035

5 Complete these multiplication pyramids. (The top number is the product of the two numbers below.)

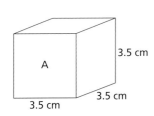

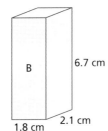

6 a) Find the volume of these boxes.

Box A: 3.5 cm, 3.5 cm, 3.5 cm

Box B: 6.7 cm, 1.8 cm, 2.1 cm

   b) Complete the statement:

      The volume of box A is .... than the volume of box B.

7 Mr Smith put 34.6 litres of petrol in his car. The cost was $1.27 per litre.

   a) Estimate how much he paid in total.   **b)** Calculate exactly how much he paid in total.

8 Apples cost $1.28 per kilogram and oranges cost $2.03 per kilogram.
Svetlana buys 1.3 kg of apples and 2.15 kg of oranges.

   a) Estimate the total cost.   **b)** Calculate the exact total cost.

9 Describe the mistake in each calculation. Correct the mistakes.

a) 5.26 × 3.4 = 178.84

$$\begin{array}{r} 5.26 \\ \times\, 3.4 \\ \hline 2104 \\ 1578 \\ \hline 178.84 \end{array}$$

b) 752 ÷ 0.16

$$16\overline{)75^{11}2} \quad \overset{4\ 7}{}$$

10 Tom is a bus driver. Last month he worked 9.5 days. Each day he worked 8.5 hours. He got paid a total of $839.80. What is his hourly rate?

11 Given the result 5.76 ÷ 0.32 = 18, deduce the results of five related calculations.

12 Estimate and then calculate:

a) 4 × (−6) × 0.5

b) −0.36 × (−2) × 5

c) −2.7 × (−0.4) × (−2)

d) (1.3 + 2.44) × 5.2

e) 6 × 3.92 ÷ 1.4

f) (8.1 − 1.74) ÷ 1.2

Thinking and working mathematically activity

In the multiplication 5.68 × 3.4 = 19.312, the total number of digits in the result (five) equals the total number of digits in the question.

Is this always true? Explore multiplications of decimal numbers with one, two or three digits. (Do not count zeros at the beginning or end of a number.)

Write a rule to describe multiplications where the result has one less digit than the question.

8.3 Multiplying and dividing by decimals between 0 and 1

Thinking and working mathematically activity

Complete a copy of the table by writing products of pairs of numbers.

×	0.8	0.9	1.0	1.1	1.2
0.8					
0.9					
1.0					
1.1					
1.2					

Describe any patterns you can see. When is the product:

- smaller than both numbers
- larger than both numbers
- smaller than one of the numbers and larger than the other?

In the multiplication $a \times b$, write rules stating when the result will be larger than a and when the result will be smaller than a.

Make a table of divisions, using the same numbers (0.8 to 1.2) as before, but writing in each space the result when you divide the left number by the top number. Round answers to two decimal places.

In the division $a \div b$, write rules stating when the result will be larger than a and when the result will be smaller than a.

Worked example 4

True or false?

a) 1.0 kg of sugar costs 90 cents. Therefore, 0.8 kg of sugar will cost more than 90 cents.

b) 0.8 m of a pipe has a mass of 240 kg. Therefore, 1.0 m of pipe will have a mass greater than 240 kg.

a) False, because $90 \times 0.8 < 90$	If you multiply a quantity by a number between 0 and 1, the result is less than the original quantity.

90 cents									
0.1 kg	0.1 kg	0.1 kg	0.1 kg	0.1 kg	0.1 kg	0.1 kg	0.1 kg	0.1 kg	0.1 kg

0.8×90 cents									
0.1 kg	0.1 kg	0.1 kg	0.1 kg	0.1 kg	0.1 kg	0.1 kg	0.1 kg		

b) True, because $240 \div 0.8 > 240$	If you divide a quantity by a number between 0 and 1, the result is greater than the original quantity.

$$240 \text{ kg} \div 0.8 = 240 \text{ kg} \div \frac{4}{5}$$
$$= 240 \text{ kg} \times \frac{5}{4}$$

240 kg			
60 kg	60 kg	60 kg	60 kg

$240 \text{ kg} \times \frac{5}{4}$				
60 kg	60 kg	60 kg	60 kg	60 kg

Exercise 3

1 State which calculations have an answer that is less than 20.

    a) 20×1.9 **b)** 20×0.9 **c)** 20×1.01 **d)** 20×0.67

2 State which calculations have an answer that is less than 23.

    a) $23 \div 0.7$ **b)** $23 \div 2.9$ **c)** $23 \div 1.05$ **d)** $23 \div 0.98$

3 Use > or < to make each statement true.

    a) 40×1.3 40   **b)** 40×1.01 40   **c)** 40×0.87 40

    d) $40 \div 1.6$ 40   **e)** $40 \div 0.7$ 40   **f)** $40 \div 0.99$ 40

4 State whether the following statements are true or false.

    a) $14 \times 1.2 > 14 \times 0.9$ **b)** $72 \times \frac{10}{11} > 72 \times 1\frac{5}{8}$ **c)** $13.8 \times 0.79 > 13.8 \times 1.08$

    d) $18 \div 0.03 > 18 \div 1.2$ **e)** $19 \div \frac{2}{3} > 19 \div 2\frac{1}{4}$ **f)** $52.7 \div 2.03 > 52.7 \div 0.96$

5 Use one of the star numbers to make these calculations true.

    0.67   3.52   0.78   1.51   0.04

    a) $5.3 \div \ldots\ldots > 5.3$ **b)** $0.62 \times \ldots\ldots > 0.62$

    c) $0.36 \div \ldots\ldots < 0.36$ **d)** $\ldots\ldots \times 0.22 < 0.22$

6 The missing number in the division below can be one of the numbers in boxes.

$2.4 \div \Box$ 0.08   0.98   1   0.1   2.4   2.41   0.23

    a) Without calculating, state which number gives the largest answer.

    b) Without calculating, state which number gives the smallest answer.

    c) Write all of the possible divisions in order, starting with the division with the smallest answer. Explain your reasoning.

7 These calculations are wrong. Explain how you know.

    a) $67 \times 0.795 = 67.025$ **b)** $89 \div 1.05 = 90.15$ **c)** $50 \div 0.8 = 45$

8 List six cereals which are sold in packets of less than 1 kg. Estimate the prices per kilogram. Use a calculator to calculate their prices.

9 Vocabulary question

    Copy and complete the sentences below using words from the box.

multiply	and	answer	greater
values	numbers	less	

When you .... or divide decimal .... , use the place .... of the decimals

to work out the equivalent calculation. When multiplying by a number between

0 .... 1, the .... will be .... than the original number. When dividing by

a number between 0 and 1, the answer will be .... than the original number.

The number of people at an event is approximately 4300.

a) Explain how you know that this number has not been rounded to the nearest thousand.

b) Is it possible that this number has been rounded to the nearest ten? Explain your answer.

Claire, Michel and Faustin want to write the limits on a quantity, x, that equals 17.5 after rounding to the nearest tenth. Describe the mistake each person has made.

a) Claire: b) Michel: c) Faustin:

$17.4 \leq x < 17.6$ $17.45 < x \leq 17.55$ $17.495 \leq x \leq 17.505$

d) Write down the correct limits.

3   A department store building is 25.6 m tall. It is 1.6 times higher than the car park next to it.

Estimate and then calculate the height of the car park.

Nina ran 12.8 km. Barack ran 2.4 times farther than Nina.

a) Estimate the distance Barack ran.   b) Calculate the distance Barack ran.

5 An iron pipe 6.3 m long has a mass of 7.56 kg.

a) Estimate the mass of 1.0 m of this pipe.   b) Calculate the mass of 1.0 m of this pipe.

6 Write a word problem that goes with each calculation.

a) $1.14 \times 0.3 = ?$ b) $0.476 \div 1.4 = ?$

The area of a rectangular garden is 83.7 m². The width of the garden is 4.5 m.

a) Estimate and then calculate the length of the garden.

b) A fence is put all around the garden. The total cost of the fence is $450.45.
Estimate and then calculate the cost of the fence per metre.

8   Copy the table below and write each of the calculations A to F in the correct place.

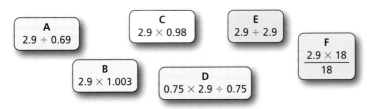

A	C	E	F
$2.9 \div 0.69$	2.9×0.98	$2.9 \div 2.9$	$\dfrac{2.9 \times 18}{18}$

B	D
2.9×1.003	$0.75 \times 2.9 \div 0.75$

Less than 2.9	Equal to 2.9	Greater than 2.9

9   Without calculating, fill in the gaps to make correct statements.

a) $4.56 \div \text{............} > 4.56$ b) $82.3 \times \text{............} < 82.3$

c) $1.4 \times \text{............} > 15.7$ d) $\text{............} \div 0.81 > 4.75$

End of chapter reflection

You should know that...	You should be able to...	Such as...
If a quantity has been rounded, you can write lower and upper limits on the exact value of the quantity.	State the upper and lower limits on numbers rounded to the nearest 1000, 100, 10, 1, 0.1 or 0.01	Write an inequality showing the limits on x if x is: **a)** 1600 to the nearest hundred **b)** 46 to the nearest whole number **c)** 0.6 to 1 decimal place **d)** 4.52 to 3 significant figures
To multiply two decimals, you can: • ignore the decimal points and multiply as integers • then write the decimal point at the correct place in the answer.	Multiply two decimals, where both decimals have at least two significant figures.	Find: **a)** 0.132×4.7 **b)** $3.7 \times (-6.8)$
To divide by a decimal, you can: • multiply the dividend and divisor by the same number, to make the divisor an integer • then use a written method of division.	Divide a positive or negative integer or decimal by a positive or negative decimal with one or two significant figures.	Find: **a)** $-686 \div (-2.8)$ **b)** $4.56 \div 1.9$
When you multiply any quantity by a number between 0 and 1, the answer is smaller than the original quantity. When you divide any quantity by a number between 0 and 1, the answer is larger than the original quantity.	Recognise the effects of multiplying and dividing by numbers between 0 and 1.	Write < or > to make each statement correct: **a)** 4.6×0.8 ..... 4.6 **b)** 3.8×1.02 ..... 3.8 **c)** $1.8 \div 0.03$ ..... 1.8 **d)** $2.9 \div 1.2$ 2.9

9 Functions and formulae

You will learn how to:
- Understand that a function is a relationship where each input has a single output.
- Generate outputs from a given function and identify inputs from a given output by considering inverse operations (including indices).
- Understand that a situation can be represented either in words or as a formula (including squares and cubes) and manipulate using knowledge of inverse operations to change the subject of a formula.

Starting point

Do you remember...

- how to represent a function as a mapping diagram, function machine or formula?

  For example, what is the formula equivalent to this function machine?

  input → | multiply by 2 | → | add 5 | → output

- how to derive and use simple formulae?

  For example, if a pack of doughnuts costs $2, what is the formula for the cost, c, of d packs of doughnuts?

- how to substitute positive and negative integers into formulae and expressions?

  For example, when $x = -4$, what is the value of $3x^2 + 4$?

This will also be helpful when...

- you learn to solve problems using inverse functions
- you learn to change the subject or rearrange complex formulae
- you learn to solve quadratic equations using the quadratic formula.

9.0 Getting started

This is a game for two people; you need a copy of the function machine, a set of number cards, dice and a set of operation cards.

Function machine

input → [|] → output

Number cards

| 0 | 1 | 2 | 3 | 4 |
| 5 | 6 | 7 | 8 | 9 |

Operation cards

| + | − | + | − | + | − |
| + | − | + | − | + | − |

Turn over the number cards and the operation cards.

First choose an operation card and place it into the first box in the function machine.

Then choose a number card and place it into the second box in the function machine.

Here is an example:

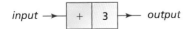

input → | + | 3 | → output

You choose a number as the output, and your partner has to calculate the input.

For example, if you choose an output of 14,

your partner would then need to work the correct input.

Always check your partner's answer by putting the input through the function machine to check that you do get the correct output.

Now swap roles, with your partner choosing the cards and you needing to work out the input.

Once you are comfortable using this simple function machine, try a double function machine, and these operation cards:

input → ▢▢ → ▢▢ → output

+ − × + −

× + − ×

9.1 Functions

Key terms

A **function** can be represented by a function machine.

The function machine for $x \rightarrow 4x + 1$ is:

x → [× 4] → $4x$ → [+ 1] → $4x + 1$

The **inverse** function can be founded by reversing the function machine.

$\dfrac{x - 1}{4}$ ← [÷ 4] ← $x - 1$ ← [− 1] ← x

So the inverse function of $x \rightarrow 4x + 1$ is $x \rightarrow \dfrac{x - 1}{4}$

Worked example 1

a) Complete the table of values for the function machine:

input → [square] → [subtract 4] → output

input	output
−4	
−2	
0	
3	
4	

b) A function is represented by $x \rightarrow x^2 + 4$. What is the inverse function?

c) If the output of the function in part **b** is 13, use the inverse function to find the input value.

a)

input	output
−4	12
−2	0
0	−4
3	5
4	12

Take each number in turn, square it, then subtract 4:

$(-4)^2 - 4 = 16 - 4 = 12$

$(-2)^2 - 4 = 4 - 4 = 0$

$(0)^2 - 4 = 0 - 4 = -4$

$(3)^2 - 4 = 9 - 4 = 5$

$(4)^2 - 4 = 16 - 4 = 12$

b) $x \rightarrow x^2 + 4$

So, the inverse function is;

$x \rightarrow \sqrt{x - 4}$

When you are finding the inverse function, you usually apply the order of operations (BIDMAS) in reverse.

Finding a square root is the inverse of squaring a number.

c) $x^2 + 4 \rightarrow 13$

inverse function:

$x \rightarrow \sqrt{x - 4}$

$13 \rightarrow \sqrt{13 - 4}$

$\rightarrow \sqrt{9}$

$\rightarrow 3$

An input gives an output of 13.

Substituting 13 into the inverse function gives an output of 3, so that must be the original value.

For the inverse to be a function, you always take the positive square root.

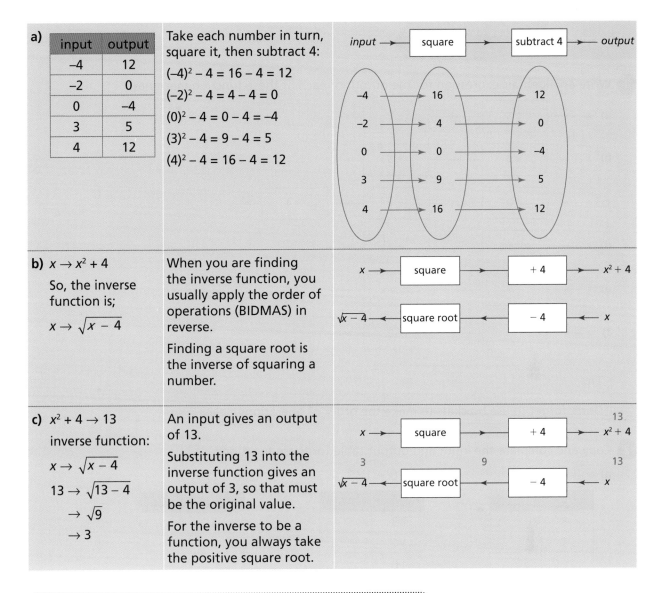

Think about

Is the order of the function machine important? For example, is $x \rightarrow 4x + 5$ the same as $x \rightarrow 5 + 4x$? How does this affect the inverse function?

1 Write an algebraic expression for the output of each of these function machines.

a) $x \longrightarrow$ [+ 5] $\longrightarrow$ [× 3] $\longrightarrow$

b) $x \longrightarrow$ [× 6] $\longrightarrow$ [+ 5] $\longrightarrow$

c) $x \longrightarrow$ [× 3] $\longrightarrow$ [− 2] $\longrightarrow$ [× 4] $\longrightarrow$

d) $x \longrightarrow$ [square] $\longrightarrow$ [+ 7] $\longrightarrow$

e) $x \longrightarrow$ [+ 7] $\longrightarrow$ [square] $\longrightarrow$

f) $x \longrightarrow$ [− 5] $\longrightarrow$ [÷ 3] $\longrightarrow$

> **Tip**
> Don't forget to use brackets to make the order of operations clear.

2 Copy and complete the input and output table for these functions.

a) $x \rightarrow 4x^2$

input	output
−3	
−1	
$\frac{1}{2}$	
3	
5	

b) $x \rightarrow (x - 1)^2$

input	output
−3	
−1	
$\frac{1}{2}$	
3	
5	

c) $x \rightarrow (x + 1)^3$

input	output
−3	
	1
$\frac{1}{2}$	
	27
5	

3 a) Write a function machine for each formula:

  i) $y = \dfrac{x+4}{5}$ ii) $y = 2x^2 - 5$ iii) $y = (x \rightarrow 1)^3$ iv) $y = \sqrt{x+6}$

b) Use your function machines to find the inverse for each function in part **a**.

4 Copy and complete this input and output table for each of the functions in question 3.

Check your answers using your inverse function machines.

If the answer is a decimal, write it correct to 2 decimal places.

input	output
−2	
0.5	
	0
4	
10	

5 Write a function machine that will give the results shown in each mapping diagram.

a)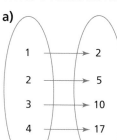

1	→	2
2	→	5
3	→	10
4	→	17

b)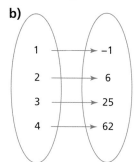

1	→	−1
2	→	6
3	→	25
4	→	62

c)

$\frac{1}{2}$	→	2
1	→	1
2	→	$\frac{1}{2}$
3	→	$\frac{1}{3}$

6 a) Copy and complete this input and output table for the function $x \to \sqrt{9x}$

input	output
1	
4	
9	
49	
100	

b) Find what input numbers would give these outputs:

i) 6 **ii)** 12 **iii)** 15

7 Jack and Tom put values into the function $y = 5x^2 - 6$.

Jack puts in the value $x = 4$ and gets the answer 74.

Tom puts in the value $x = -4$ and also gets the answer 74.

a) Who is correct?

b) What is the inverse of the function $y = 5x^2 - 6$?

c) If you put the value 74 into the inverse function, what value would you get out?

Explain how you know.

8 a) Copy and complete the input and output table for the function $x \to \dfrac{1}{x}$

b) What do you notice about the answers?

input	output
0.2	
0.5	
	1
2	
5	

Discuss

For the function $x \to \sqrt{x}$, you have to limit the values of x you can use to make the function work.

For the function $x \to \sqrt[3]{x}$, you do not have to limit the values of x. Why is that? Are there any other functions in the exercise where there are limited values of x you can use?

When the inverse function is the same as the original function, it is called a self-inverse function.

In other words, if you apply the function twice, you get back to your starting value every time.

Here are some functions.

$x \rightarrow 10 - x$ $x \rightarrow \dfrac{1}{x^2}$ $x \rightarrow 10 + x$ $x \rightarrow \dfrac{1}{x}$

$x \rightarrow \dfrac{5}{x}$ $x \rightarrow 3 - 2x$ $x \rightarrow 3 - \dfrac{1}{x}$ $x \rightarrow 1 - x$ $x \rightarrow \dfrac{x}{5}$

Try some different input values for the functions. Include fractions and negative numbers.

Which of the functions are self-inverse? Show how you know.

Can you find any other self-inverse functions?

9.2 Formulae

Key terms

A **variable** is a quantity that can change. We represent variables with letters of the alphabet.

A **formula** is a mathematical relationship between two or more variables expressed algebraically.

A formula does not mean anything unless you say what your variables represent.

Changing the subject of a formula is when you rearrange your formula so that a different variable can be calculated.

Worked example 2

a) Make x the subject of the formula $A = 3x^2 - 4$

b) Express the area of this compound shape in terms of x.

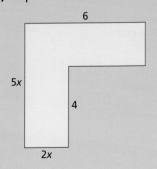

a)

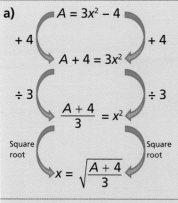

$$A = 3x^2 - 4$$

$+4 \qquad +4$

$$A + 4 = 3x^2$$

$\div 3 \qquad \div 3$

$$\frac{A + 4}{3} = x^2$$

Square root   Square root

$$x = \sqrt{\frac{A + 4}{3}}$$

'Make x the subject of the formula' means that you want to rearrange it to get $x = ...$

Think about what you have done to x to get A, then reverse the operations, and the order of operations to get back to your starting value.

Formula:

$x \mapsto$ square $\mapsto$ multiply by 3 $\mapsto$ subtract $4 \mapsto A$

Rearranging:

$A \mapsto$ add $4 \mapsto$ divide by 3 $\mapsto$ square root $\mapsto x$

b) $A = 6(5x - 4) + 4 \times 2x$

$\quad A = 30x - 24 + 8x$

$\quad A = 38x - 24$

First calculate the missing lengths in terms of x.

Then split the shape into rectangles.

Find the area of each rectangle by multiplying the side lengths.

Write your answer as simply as possible.

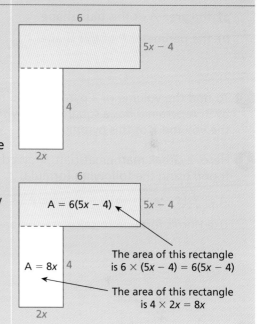

The area of this rectangle is $6 \times (5x - 4) = 6(5x - 4)$

The area of this rectangle is $4 \times 2x = 8x$

> **Think about**
>
> Could you have split this shape into different rectangles? Would you still get the same answer?

1 Write a formula to find:

   a) the area A of this green shape

   b) the perimeter P of this shape.

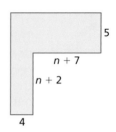

2 Write a formula to find:

   a) the area A of this blue shape

   b) the perimeter P of this shape.

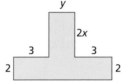

3 To find the volume of a pyramid, you multiply the height by the area of the base and then divide by 3. A pyramid has a square base, with sides of length x cm, and a height h. Find a formula for the volume V of this pyramid.

4 Hero, a Greek mathematician, showed that the area A of a triangle with sides a, b and c can be found using the following formula:

$$A = \sqrt{s(s-a)(s-b)(s-c)} \quad \text{where } s = \frac{1}{2}(a+b+c)$$

Use Hero's formula to calculate the area of this triangle:

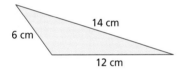

5 $A = \frac{1}{2}h(a+b)$ is the formula for the area of a trapezium.

   a) Make h the subject of the formula.

   b) Make a the subject of the formula.

6 **a)** Make Q the subject of $V = \dfrac{2Q + W}{3}$

   b) Make u the subject of $v^2 = u^2 - 2as$

7 Match each formula with its equivalent when p is made the subject.

$q = (p+4)^2$	$q = (5p-3)^2$	$p = \dfrac{\sqrt{q}+3}{5}$	$p = \dfrac{q - r^2}{m}$
$q = \dfrac{p^2 - 3}{5}$	$q = \dfrac{mp}{r^2}$	$p = \sqrt{5q+3}$	$p = \dfrac{qr^2}{m}$
$q = mp + r^2$	$q = 5p^2 - 3$	$p = \sqrt{\dfrac{q+3}{5}}$	$p = \sqrt{q} - 4$

8 Rearrange each of the following to make t the subject:

a) $y = t^3$ **b)** $y = 2\sqrt{t}$ **c)** $y = \dfrac{4(t - 2)}{5}$ **d)** $y = \dfrac{2t}{9} - 3$

e) $y = 5 - 2t$ **f)** $y = 7(t - 1)^2$ **g)** $y = 4at + 2b^2$ **h)** $y = \dfrac{a}{t}$

9 Henri, André and Louis rearrange the formula $y = \dfrac{(x + 3)^2}{5}$ to make x the subject.

Henri thinks the formula will become $x = \sqrt{5y - 3}$

André thinks the formula will become $x = 5y^2 - 3$

Louis thinks the formula will become $\sqrt{5y} - 3$

Who is correct?

Explain how you know. What mistakes have the other two people made?

10 $P = \pi r + 2r$ is the formula for the perimeter of a semicircle with radius r.

Make r the subject of the formula.

> **Tip**
>
> Factorise this formula so that $P = r (.... + ...)$

11 Make r the subject of each formula.

a) $S = 3rg - 2h$ **b)** $T = \dfrac{\pi r - p}{q}$ **c)** $V = \dfrac{1}{3}\pi rl$ **d)** $A = x^2 + ry$

Thinking and working mathematically activity

Technology question A box is made from a piece of cardboard measuring 18 cm by 20 cm, with squares cut from the corners before folding up the sides to make a box.

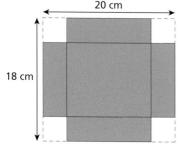

a) Write down a formula for the volume V of the box, with squares of side length x cut from the corners.

b) Use spreadsheet software to work out what size square should be cut from each corner to make a volume of:

i) 400 cm³   **ii)** the largest possible volume

Explain how you can be sure that you have found the largest possible volume.

> **Tip**
>
> You may need to use non-integer values for the size of the squares.
>
> Explain how you can be sure that you have found the largest possible volume.

1 Write a function machine for each of these functions.

a) $x \to 2x - 5$ **b)** $x \to 2(x - 5)$ **c)** $x \to x^2 + 3$ **d)** $x \to \left(\dfrac{x}{2}\right)^2$

2 Match each function with the correct mapping diagram.

a) $x \to \dfrac{2}{x}$ **b)** $x \to (x - 1)^3$ **c)** $x \to (2 - x)^2$

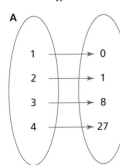
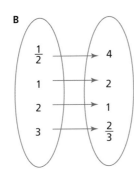
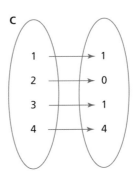

3 Copy and complete the input and output tables for each function.

a) $x \to 2x^2 + 1$ **b)** $x \to \sqrt{4x}$ **c)** $x \to 100 - x^2$

input	output
−2	
0.5	
	3
	9
5	
	163

input	output
0.25	
1	
	4
9	
	8
	10

input	output
−6	
−1	
	100
	36
12	
	−300

4 Find the inverse function for each of the functions in question 3.

5 a) Write the formula to find:

  i) the area A of this shape.

  ii) the perimeter P of this shape.

  b) Find the area and perimeter of this shape when $m = 4$. (All lengths are in cm.)

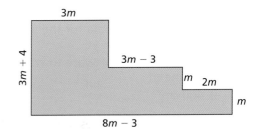

6 The formula for the volume of this triangular prism is $V = \dfrac{1}{2} bhL$

  a) Rearrange the formula to make L the subject.

  The ends of this prism are equilateral triangles.

  b) Write down, in its simplest form, a formula for the surface area A of this prism.

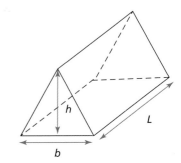

7 Rearrange each of these to make m the subject.

  a) $3(2m - 3) = n$ **b)** $p(m + 1) = q$ **c)** $\dfrac{m + b}{5} = c$

  d) $\dfrac{m}{n} + 7 = w$ **e)** $\sqrt{m} - 2 = n$ **f)** $\sqrt{m + a} = b$

8 Huw and Kelly rearrange the formula $a^2x - 9 = b$ to make x the subject.

Huw says the answer is $x = b - a^2 + 9$

Kelly says the answer is $x = \dfrac{b + 9}{a^2}$

Who is correct? Explain how you know.

End of chapter reflection

You should know that...	You should be able to...	Such as...
Functions can be formed using indices and roots.	Find output values for functions involving indices and roots.	Complete a table of values for the function $x \rightarrow 3x^2 - 1$, with integer values of x from -3 to 3
The inverse of a function can be found by reversing the function machine.	Find the inverse function of a function involving an index number or root.	Find the inverse function of: **a)** $x \rightarrow \sqrt{x} - 7$ **b)** $x \rightarrow \dfrac{x^2}{2}$
A formula can be written to connect variables. Many formulae are used in mathematics and other subjects.	Write a formula for a given situation.	Write a formula to calculate the area of this shape.
When substituting a value into an expression or a formula, you must always apply the correct order of operations.	Substitute values into a formula and correctly calculate the result.	$n = am^2 + 5$ Find the value of n when $a = 0.4$ and $m = -6$
When rearranging a formula to change the subject, you need to pay close attention to the order of operations.	Rearrange a formula to make a given term the subject.	Make m the subject of the formula $n = am^2 + 5$

You will learn how to:

- Construct 60°, 45° and 30° angles and regular polygons.
- Use knowledge of bearings and scaling to interpret position on maps and plans.
- Use knowledge of coordinates to find points on a line segment.

Starting point

Do you remember...

- how to construct a perpendicular line bisector? For example:

  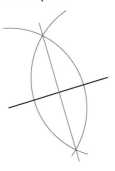

- how to construct an angle bisector? For example:

  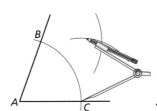 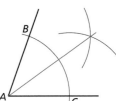

- what a bearing is?

  For example, write down the bearing of the compass direction of south east.

- how to find the midpoint of a line segment?

  For example, find the coordinates of the midpoint of the line segment joining (5, –2) and (11, 8).

This will also be helpful when...

- you learn to solve problems about loci and bearings
- you use maps in real life.

10.0 Getting started

Octagon design

- Draw a large square on a piece of squared paper.
- Draw the diagonals of the square.

- Open your compass to half of one of the square's diagonals. Use the compass to draw an arc with the centre on one vertex of the square.

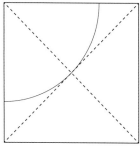

- Repeat at the other three vertices of the square.

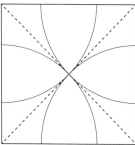

- Join the points where the arcs meet the sides of the square to make an octagon.

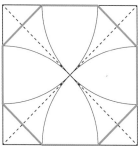

10.1 Construction

Key terms

To **construct** means to draw accurately. In this chapter it means drawing without using a protractor.

When you **inscribe** a polygon in a circle, you draw the polygon so that its vertices all touch the circumference of the circle.

Constructing angles

Construct 90°

An angle of 90° is created when you construct a perpendicular bisector.

Construct 45°

This is half of 90°.

First construct an angle of 90°, then bisect the angle.

Construct 60°

Draw a line segment labelled A at one end.

Open your compass to about 3 cm. With point A as centre, draw an arc to intersect the line at point B.

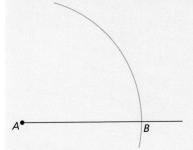

With point B as centre, draw a second arc which intersects the first arc. Use the same radius as the first arc.

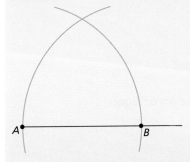

Join point A to the point where the arcs intersect, C. Angle BAC is 60°

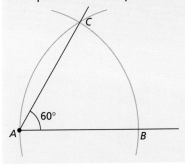

Worked example 1

Using only a straight edge and compass, inscribe an equilateral triangle in a circle.

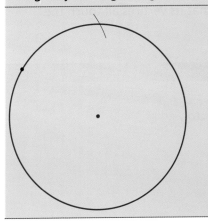

Draw a circle. Keeping the compass opening equal to the circle's radius, place the compass point on the circumference and draw a short arc cutting the circumference.

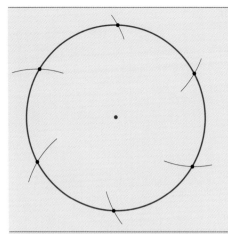

Keeping the compass opening the same, draw evenly spaced arcs around the circumference.

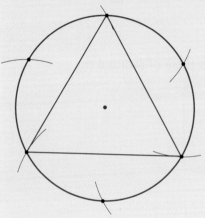

Use a straight edge to join every other point on the circumference.

Discuss

How could you adapt this method to inscribe a regular hexagon in a circle?

Exercise 1

1. **a)** Construct an angle of 90°.

   b) Bisect this angle to construct an angle of 45°.

2. Inscribe the following shapes in a circle. Make the radius of the circle at least 5 cm.

   a) equilateral triangle   **b)** regular hexagon

3. Follow these steps to inscribe a square in a circle.

   • Draw a circle with radius about 4 cm.

   • Draw in the diameter of the circle. Label the diameter *AB*.

   • Construct another diameter perpendicular to the first one. Label this diameter *CD*.

   • Join the four points to make a quadrilateral *ACBD*.

4 Construct these triangles accurately without using a protractor.

a)

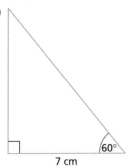

7 cm

b)

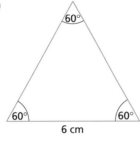

6 cm

5 Construct these angles. Check your accuracy using a protractor.

a) 30°   b) 15°   c) 22.5°

6 Construct an inscribed square with diagonals of length 10 cm.

7 Using a ruler and a compass only, inscribe a regular octagon in a circle of radius 6 cm.

> **Tip**
>
> Begin by inscribing a square inside the circle. Then construct the perpendicular bisectors of the sides of the square to find the remaining vertices of the octagon.

8 Using a ruler and compass only, construct these triangles.

a)

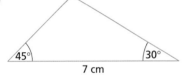

7 cm

b)

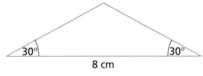

8 cm

9 Toby's attempt at constructing a hexagon inside a circle shown below. Describe the mistake he has made.

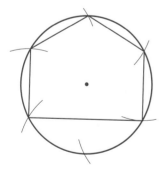

10 Construct an angle of 75°

11 Greta is asked to construct an isosceles triangle with an angle of 135°.

She says, 'It can't be done, because I can't construct an angle of 135°.'

Is Greta correct? Explain how you know.

Thinking and working mathematically activity

When you inscribe an equilateral triangle or a regular hexagon in a circle, you make six evenly spaced points around the circumference. List the other shapes you could make by joining some of these points.

When you inscribe a regular octagon in a circle, you bisect two of the 90° angles made by the perpendicular diameters. If you bisect the angles more than once, what regular polygons can you construct?

Try to inscribe a regular dodecagon (a 12-sided shape) in a circle.

There are many websites where you can use a virtual ruler and compass for practising construction methods.

10.2 Bearings

Key terms

A **bearing** is an angle measured clockwise from north.

A **scale drawing** is a reproduction of a drawing but with all the sides increased or decreased in size in the same ratio.

The **scale** on a map is given as a ratio.
The scale 1 : 100 000 means that 1 cm on the map represents 100 000 cm (1 km) in real life.

> **Did you know?**
>
> The science of drawing maps is called cartography.

Worked example 2

The diagram shows the position of two mountain tops, *A* and *B*.

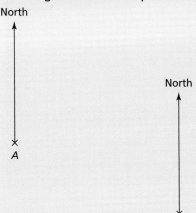

B is 8 km from *A* on a bearing of 120°.

A third mountain *C* is on a bearing of 140° from *A* and on a bearing of 270° from *B*.

a) Use a scale of 1 : 200 000 to make a scale drawing of the diagram.

b) Mark the position of the mountain top *C* on the scale drawing.

c) Find the distance between *B* and *C* in real life.

a)

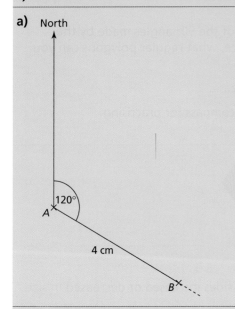

The scale 1 : 200 000 means that 1 cm represents

200 000 cm = 2000 m

= 2 km

B is 8 km from *A* in real life, so the distance between *B* and *A* on the map needs to be 4 cm.

Mark the point *A* and draw in a North line. Measure a bearing of 120° from *A* and draw a line from *A*, extending along the 120° direction.

Measure a distance of 4 cm from *A* along this line and mark the point as *B*.

b)

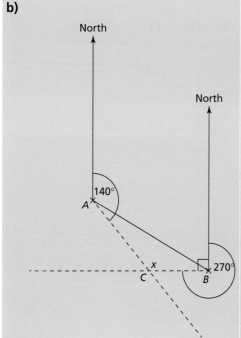

Draw the North line at *B*.

Measure a bearing of 270° from *B*.

Measure a bearing of 140° from *A*.

Mark *C* at the point where the lines intersect.

c) The distance between B and C is 1.8 cm

The scale is: 1 cm represents 2 km

So, the real-life distance is $1.8 \times 2 = 3.6$ km.

Measure the distance from B to C on the map.

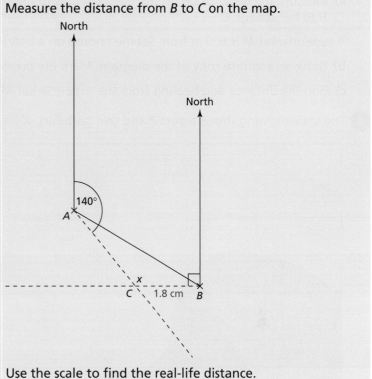

Use the scale to find the real-life distance.

Exercise 2

1 A map is drawn to a scale of 1 : 400 000

    a) A road measures 6.7 cm on the map. Work out the length (in kilometres) of the actual road.

    b) The distance between two towns is 18 km. How far apart will they be on the map?

2 *A is 5 km due north from C. B is due east from A.*

    B is on a bearing of 045° from C.

    a) Draw a scale drawing of the three points. Use the scale 1 : 100 000

    b) How far is B from C in real life?

3 The scale drawing shows the position of Sabine's house H and her college C.

    C is on a bearing of 280° from H.

    The scale of the map is 1 : 20 000.

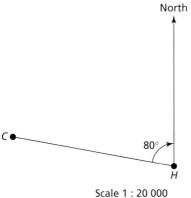

Scale 1 : 20 000

a) Measure the distance *CH* on the map. Use this to find the actual distance from Sabine's house *H* to her college *C*.

A supermarket *M* is 600 m from Sabine's house on a bearing of 150°.

b) Draw an accurate copy of the diagram. Mark the position of *M* on your diagram.

c) Find the distance and bearing from the supermarket *M* to Sabine's college *C*.

4 The scale drawing shows a port *P* and two harbours, *X* and *Y*. The scale is 1 : 50 000

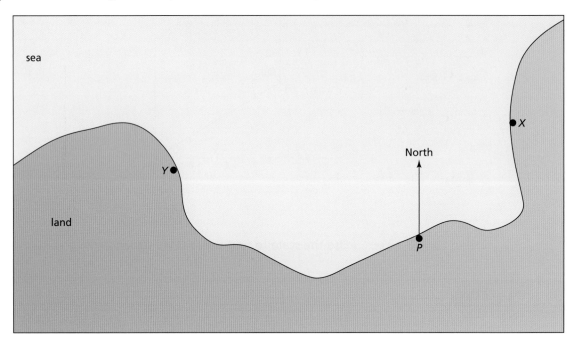

a) Measure the bearings of *X* and *Y* from *P*.

b) Find the distance of *Y* from *P* in real life.

A boat asks for help. It is at a point *B* which is 065° from *Y* and 290° from *X*.

c) Draw a copy of the diagram. Mark the position of *B* on your diagram.

d) Find the real-life distance from the port *P* to the boat *B*.

5 The diagram shows three villages, *A*, *B* and *C*.

a) Draw a scale drawing of the diagram.
   Use the scale 1 : 200 000

b) A fourth village *D* is on a bearing of 278° from *C* and on a bearing of 115° from *A*. Mark the position of *D* on your scale drawing.

c) Find how far village *D* is from *A*.

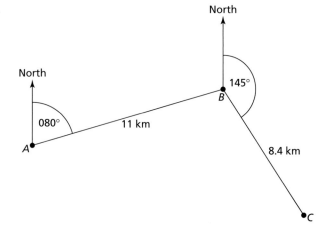

6 Harry leaves his house *H* and goes for a run.

He first runs 5.8 km on a bearing of 038° to point *P*.

He then runs 7.5 km on a bearing of 170° to point *Q*.

a) Draw a scale drawing to show Harry's journey. Use a scale of 1 : 100 000.

Harry finally runs from *Q* straight home.

b) Find the bearing he runs travelling home from *Q*.

c) Harry says he ran a total distance of more than 20 km. Is he correct? Show how you found your answer.

Thinking and working mathematically activity

Technology question *ABCDEFGHI* is a regular nonagon.

A is the topmost point of the polygon.

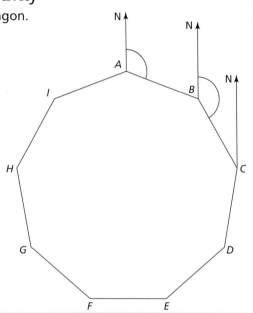

- Draw *ABCDEFGHI* using dynamic geometry software.

Investigate this sequence of bearings: *B* from *A*, *C* from *B*, *D* from *C*, …

- Now begin with different regular polygons.

Investigate the same sequence of bearings.

- Try to explain any patterns you notice.

10.3 Points on a line segment

Key terms

A **line segment** is a section of a line. It has two end points.

Thinking and working mathematically activity

Draw a pair of axes and draw three line segments:

AB   from *A*(1, 5) to *B*(5, 9)

CD   from *C*(3, 2) to *D*(9, 5)

EF   from *E*(2, 1) to *F*(10, 5)

- Find the coordinates of the point $\frac{1}{2}$ of the way along *AB*.

- Find the coordinates of the point $\frac{1}{3}$ of the way along *CD* from *C*.

- Find the coordinates of the point $\frac{1}{4}$ of the way along *EF* from *E*.

- What do you notice about these sets of coordinates and the fraction?

- Write a rule to predict the coordinates of a fraction part along a line segment.

Worked example 3

A and *B* are the points with coordinates (–5, 2) and (4, –1)

Find the coordinates of the point *C* that is one third of the way along *AB* from *A*.

Difference in *x*-coordinates is 4 – (–5) = 9	To get from *A* to *B* you go	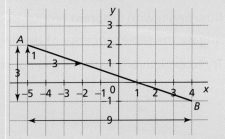
Difference in *y*-coordinates is –1 – 2 = –3	• 9 squares right	
	• 3 squares down.	
x-coordinate of *C* is	Find one third of these differences:	
$-5 + \frac{1}{3} \times 9 = -2$	One third of 9 is 3	
y-coordinate of *C* is	One third of 3 is 1.	
$2 + \frac{1}{3} \times (-3) = 1$	To find the coordinates of *C* you need to go:	
So the coordinates of *C* are (–2, 1)	• 3 squares right from *A*	
	• 1 square down from *A*.	
	This gives the coordinates:	
	(–5 + 3, 2 – 1) or (–2, 1)	

1 Write down the coordinates of the point:

a) half way along *AB*

b) $\frac{1}{4}$ of the way along *AB* from point *A*

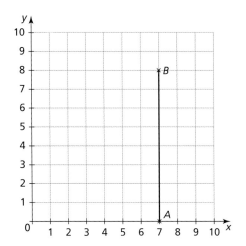

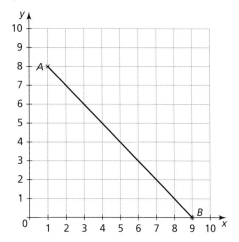

c) $\frac{1}{3}$ of the way along *AB* from point *A*

d) $\frac{1}{5}$ of the way along *AB* from point *A*

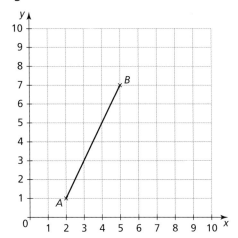

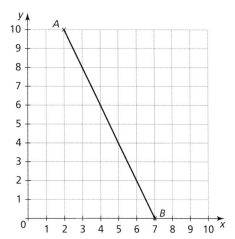

2 The diagram shows five points, *A*, *B*, *C*, *D* and *E*.

Find the coordinates of the point that lies:

a) one third of the way along line segment *EA*

b) one quarter of the way along line segment *EC*

c) one fifth of the way along line segment *DB*

d) one quarter of the way along line segment *BC*.

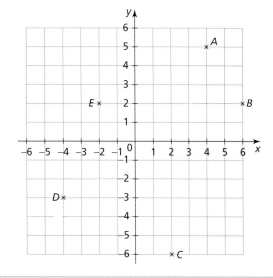

3 Find the coordinates of the point that lies:

a) one third of the way along the line segment joining (1, 7) and (10, 4)

b) one fifth of the way along the line segment joining (–7, 8) and (3, 3)

c) one eighth of the way along the line segment joining (–11, 16) and (13, 0)

d) three tenths of the way along the line segment joining (32, 6) and (–8, –4)

e) two thirds of the way along the line segment joining (1, 4) and (13, 13)

f) three fifths of the way along the line segment joining (–2, –6) and (3, 4).

4 PQRS is a parallelogram. The coordinates of P, Q and R are P(8, 1), Q(6, 7) and R(12, 9).

a) Find the coordinates of the point one quarter of the way along the diagonal PR.

b) Find the coordinates of the point one quarter of the way along the diagonal QS.

5 A has coordinates (1, 3). B has coordinates (5, –2).

B is one third of the way along line segment AC. Find the coordinates of C.

6 The origin O, point A_1, and point A_2 are equally spaced along the same line such that the distance OA_1 is equal to the distance A_1A_2. The coordinates of point A_1 are (4, 3).

a) What are the coordinates of point A_2?

b) The sequence of points A_1, A_2, A_3, A_4, …. are all equally spaced on this line.

Sophia says the coordinates of A_{26} must be (104, 78).

Is Sophia correct? Explain your answer.

c) Find the coordinates of the point one third of the way between A_4 and A_{10}.

7 The point two fifths of the way from P(4, a) to Q(b, –15) has coordinates (10, –3). Find the values of a and b.

8 The point G(5, 10) is one quarter of the way from E(a, b) to F(c, d).

a, b, c and d are all prime numbers less than 20.

Find the coordinates of the midpoint of GF.

9 ABCD is a rectangle where A is (–2, –3).

The coordinates of the midpoint of the line segment AD are (–3, –1).

The coordinates of the point one quarter of the way along AB are (0, –2).

Find the coordinates of C.

10 P and Q are points with coordinates (r, 2r – 1) and (4r + 6, 8r + 11) respectively.

S is the point one third of the way along PQ.

a) Find the coordinates of S in terms of r.

b) The coordinates of S are (14, 27). What are the coordinates of the point one quarter of the way along SQ?

> **Think about**
>
> C is one quarter of the way along the line segment AB.
>
> D is the midpoint of CB.
>
> What fraction of the way along the line segment AB is D?

1 **a)** These instructions for inscribing an octagon in a circle are in the wrong order.
Write the correct order.

A	Draw a diameter of the circle.
B	Join the eight points where the bisector lines meet the circumference.
C	Draw a circle.
D	Bisect two adjacent 90° angles at the centre of the circle, extending the bisectors to meet the circumference.
E	Construct a perpendicular bisector of the diameter, extending it to meet the circumference.

b) Gulzhan has tried to construct an octagon inscribed inside a circle. She has made a mistake. Describe the mistake she has made.

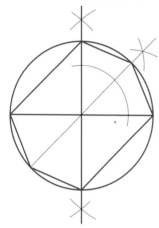

2 A regular hexagon is inscribed in a circle of radius 8 cm. Without doing the construction, find the length of one side of the hexagon. Show the steps in your reasoning.

3 A mountain rescue centre *M* is on a bearing of 105° from an airport *A*.

The distance between *A* and *M* is 36 km.

a) Draw a scale drawing to show the positions of *A* and *M*. Use the scale 1 : 500 000.

A call is received to say someone on the mountain needs help. The caller is at point *H*, on a bearing of 150° from *A* and 260° from *M*.

b) Show the position of *H* on your scale drawing.

c) A helicopter flies from *A* to *H*. Find the distance it flies.

4 Find the point that lies one quarter of the way along the line segment joining (−5, 4) to (7, −4).

5 The coordinates of a point *P* are (−2, 3). The point *R*(3, −6) is one third of the way along the line segment *PQ*. Find the coordinates of *Q*.

6 *T* and *B* are points with coordinates *T*(5, 3) and *B*(33, 19).

T is one fifth of the way along a line *AB*.

Ollie says *A* must be (−2, −1). Show that Ollie is correct.

End of chapter reflection

You should know that...	You should be able to...	Such as...
A compass can be used to construct special angles. An inscribed polygon is one constructed inside a circle.	Construct angles of 30°, 45° and 60°. Construct an inscribed equilateral triangle and square. Construct an inscribed regular hexagon and octagon.	Construct an angle of 45°. Draw an inscribed regular hexagon inside a circle with radius 4 cm.
Bearings can be used to find positions from given points.	Mark positions of points on scale drawings.	A box of treasure is located on a bearing of 060° from *A* and 330° from *B*. Mark the position of the treasure box on the map. N ↑ *A* ✕ ↑ ✕ *B*
The coordinates of a point that is a fraction of the way along a line segment can be calculated from the coordinates of the end points.	Find a point that is a fraction of the way along a given line segment.	Find the point that is one third of the way along the line segment joining (−2, −6) and (13, 6)

11 Fractions

You will learn how to:

- Deduce whether fractions will have recurring or terminating decimal equivalents.
- Estimate, add and subtract proper and improper fractions, and mixed numbers, using the order of operations.
- Estimate, multiply and divide fractions, interpret division as a multiplicative inverse, and cancel common factors before multiplying or dividing.

Starting point

Do you remember…

- how to convert a fraction to a decimal by division?

  For example, convert $\frac{3}{8}$ to a decimal.

- how to recognise whether a decimal is terminating or recurring?

  For example, convert these fractions to decimals and state which are recurring: $\frac{2}{3}, \frac{2}{5}, \frac{2}{11}$

- how to add and subtract mixed numbers?

  For example, find $4\frac{3}{5} + 1\frac{1}{3}$ and $3\frac{5}{8} - 2\frac{1}{4}$

- how to multiply and divide proper fractions?

  For example, find $\frac{8}{15} \times \frac{9}{10}$ and $\frac{5}{9} \div \frac{1}{6}$

- how to multiply a mixed number by an integer?

  For example, find $2\frac{3}{4} \times 5$

- how to divide an integer by a proper fraction?

  For example, find $5 \div \frac{3}{4}$

This will also be helpful when…

- you learn to convert recurring decimals to fractions.

11.0 Getting started

Copy and complete the table below, adding one row for each of these fractions.

$\frac{1}{2}$ $\frac{1}{3}$ $\frac{1}{4}$ $\frac{1}{5}$ $\frac{1}{6}$ $\frac{1}{7}$ $\frac{1}{8}$ $\frac{1}{9}$ $\frac{1}{10}$ $\frac{1}{11}$ $\frac{1}{12}$

$\frac{1}{14}$ $\frac{1}{15}$ $\frac{1}{16}$ $\frac{1}{18}$ $\frac{1}{20}$ $\frac{1}{21}$ $\frac{1}{25}$ $\frac{1}{30}$ $\frac{1}{35}$ $\frac{1}{40}$ $\frac{1}{50}$

Fraction	Prime factors of the denominator	Decimal equivalent	Terminating or recurring?
$\frac{1}{2}$			

Write a rule for predicting which fractions have recurring decimal equivalents.

11.1 Recurring and terminating decimals

Key terms

A **terminating decimal** is a decimal that stops after a number of decimal places, for example 0.6 or 0.74 or 0.31406

A **recurring decimal** has digits that continue forever in a repeating pattern. We write dots or a bar above the part that repeats. For example, $0.5\dot{6}\dot{3}$ means that the 63 repeats: 0.5636363...

To see if a fraction is equivalent to a recurring decimal, look at the prime factors of the denominator. If all of the prime factors are 2s and/or 5s, then it is a terminating decimal. If there are any other prime factors (for example, 3, 7, 11), then it is a recurring decimal.

> **Did you know?**
>
> The fraction $\dfrac{1}{9^2} = 0.\dot{0}12345678\dot{9}$. Explore the decimal equivalents of other fractions with 1 as the numerator and a power of 9 as the denominator.

Worked example 1

a) If $\dfrac{1}{9} = 0.\dot{1}$, write $\dfrac{5}{9}$ as a decimal.

b) From the list below, write the fractions that are equivalent to recurring decimals.

$\dfrac{5}{14}$ $\dfrac{3}{15}$ $\dfrac{8}{15}$ $\dfrac{3}{20}$

a) $\dfrac{5}{9} = 5 \times \dfrac{1}{9}$ $= 5 \times 0.\dot{1}$ $= 0.\dot{5}$	If you multiply $\dfrac{1}{9} = 0.111\dots$ by 5, you get 0.555 ...
b) $\dfrac{5}{14}$ $14 = 2 \times 7$ $\dfrac{3}{15} = \dfrac{1}{5}$ $5 = 5$ $\dfrac{8}{15}$ $15 = 3 \times 5$ $\dfrac{3}{20}$ $20 = 2 \times 2 \times 5$ $\dfrac{1}{5} = \dfrac{2}{10} = 0.2$ $\dfrac{3}{20} = \dfrac{15}{100} = 0.15$ $\dfrac{5}{14}$ and $\dfrac{8}{15}$ are equivalent to recurring decimals.	With the fraction in its simplest form, look at the prime factors of the denominator. If there are any prime factors that are *not* 2 or 5, then the decimal recurs. If the prime factors are only 2s and/or 5s, then the decimal terminates.

2 – 6

Use a calculator to convert each fraction to a decimal.
If the decimal recurs, write it using dot notation.

a) $\frac{7}{9}$ b) $\frac{1}{6}$ c) $\frac{7}{8}$ d) $\frac{5}{12}$ e) $\frac{13}{22}$

2 Without division calculations, write the fractions that have recurring decimal equivalents.

$\frac{3}{8}$ $\frac{2}{3}$ $\frac{7}{10}$ $\frac{37}{100}$ $\frac{4}{11}$ $\frac{3}{5}$ $\frac{2}{15}$ $\frac{9}{16}$

Check your answers by using division to convert each fraction to a decimal.

3 The fractions below are in their simplest forms.

$\frac{29}{80}$ $\frac{53}{60}$ $\frac{13}{32}$ $\frac{37}{50}$ $\frac{11}{40}$ $\frac{41}{75}$ $\frac{23}{56}$

a) Write the prime factors of the denominator of each fraction.

b) Write the fractions that have recurring decimal equivalents.

Tedros says that the decimal equivalent of $\frac{21}{56}$ is recurring, because one of the prime factors of 56 is 7.

Is he correct? Explain your answer.

5 a) Use division to find the decimal equivalent of $\frac{1}{9}$.

b) Without using division, write the decimal equivalent of $\frac{2}{9}$. Explain your reasoning.

c) Use a fraction multiplication to show that $0.\dot{9} = 1$

6 Consider fractions with denominators between 1 and 101, written in their simplest form.

For which denominators do the decimal equivalents terminate?

> **Tip**
> There are 14 possible values.

Thinking and working mathematically activity

Below is a table of fractions with powers of 5 as denominators, and equivalent fractions with powers of 10 as denominators.
On a copy of the table, fill in the gaps.

Fraction with power of 5 denominator	Fraction with denominator multiplied out	Multiplier	Equivalent fraction with power of 10 denominator
$\frac{1}{5^1} =$	$\frac{1}{5} =$	$\times \frac{2}{2} =$	$\frac{2}{10}$
$\frac{1}{5^2} =$			
$\frac{1}{5^3} =$			
$\frac{1}{5^4} =$			

Describe any pattern you can see. Explain why fractions $\frac{1}{5^n}$, where n is a positive integer, have decimal equivalents that terminate.

Make a similar table, showing fractions $\frac{1}{2^1}$, $\frac{1}{2^2}$, $\frac{1}{2^3}$ and $\frac{1}{2^4}$. Fill in the table.

Describe any pattern you can see. Explain why fractions $\frac{1}{2^n}$, where n is a positive integer, have decimal equivalents that terminate.

11.2 Adding and subtracting fractions

Worked example 2

Estimate and then calculate $3\frac{3}{5} - \left(\frac{2}{3} + 1\frac{4}{15}\right)$

$3\frac{3}{5} - \left(\frac{2}{3} + 1\frac{4}{15}\right)$ $\approx 4 - (1 + 1)$ $= 2$	To estimate, round each mixed number to the nearest integer.	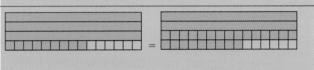
$3\frac{3}{5} - \left(\frac{2}{3} + 1\frac{4}{15}\right)$ $= 3\frac{9}{15} - \left(\frac{10}{15} + 1\frac{4}{15}\right)$	Write the fractions with a common denominator.	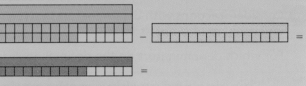
$= 3\frac{9}{15} - 1\frac{14}{15}$	Follow the usual order of operations. Add the fractions in brackets first.	
$= 2\frac{24}{15} - 1\frac{14}{15}$	$\frac{9}{15}$ is smaller than $\frac{14}{15}$, so change 1 in the first mixed number into fifteenths.	
$= 1\frac{10}{15}$ $= 1\frac{2}{3}$	Do the integer and fraction subtractions separately. Then simplify.	

> **Think about**
>
> The method shown in Worked example 2 is not the only possible method.
> Write two other methods. Comment on which method you prefer.

Exercise 2   1 – 6

1 Estimate and then calculate:

a) $5\frac{3}{8} + 1\frac{1}{4}$

b) $3\frac{1}{3} - \frac{13}{8}$

c) $\frac{11}{5} - 1\frac{3}{10}$

d) $\frac{17}{3} - \frac{5}{6} + 2\frac{3}{8}$

e) $1\frac{1}{5} - \frac{3}{4} - \frac{1}{10}$

f) $3\frac{5}{16} - \frac{9}{8} + \frac{13}{4}$

g) $3\frac{1}{6} + 2\frac{2}{9} - \frac{3}{4}$

h) $4\frac{5}{9} - 1\frac{1}{3} - 2\frac{7}{18}$

i) $2\frac{7}{8} - 1\frac{3}{4} + 1\frac{5}{6}$

2 The table shows the times that Amy trained last week.

Monday	Wednesday	Saturday
$2\frac{2}{3}$ hours	$1\frac{1}{4}$ hours	$2\frac{1}{6}$ hours

a) How many hours did Amy train during the week?

b) Amy trained a total of $4\frac{3}{5}$ hours on Saturday and Sunday.
How many hours did she train on Sunday?

3 Which expression is larger? Find how much larger.

A $2\frac{2}{3} + 3\frac{5}{6} + \frac{3}{4}$ **B** $7\frac{7}{8} - \left(1\frac{5}{12} + 1\frac{1}{3}\right)$

▶ The perimeter is the same length for both of these triangles. Find the missing length.

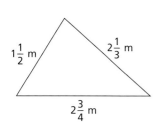
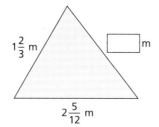

> **Tip**
>
> Start by finding the perimeter of the first triangle.

First triangle: $1\frac{1}{2}$ m, $2\frac{1}{3}$ m, $2\frac{3}{4}$ m

Second triangle: $1\frac{2}{3}$ m, ☐ m, $2\frac{5}{12}$ m

5 Calculate these:

a) $\frac{7}{12} + \left(\frac{5}{6} - \frac{1}{3}\right)$ **b)** $\frac{14}{15} - \left(\frac{1}{3} + \frac{3}{10}\right)$ **c)** $\frac{9}{16} - \left(\frac{5}{8} - \frac{1}{4}\right)$

6 Estimate and then calculate each answer.

If an answer is greater than 1, write it as a mixed number.

a) $7 - \left(\frac{2}{3} + \frac{11}{6} + \frac{3}{4}\right)$ **b)** $5\frac{1}{2} - \left(4\frac{7}{8} - 1\frac{1}{3}\right)$ **c)** $6\frac{7}{8} - \left(2\frac{1}{4} + 1\frac{1}{12}\right)$

d) $\left(2\frac{1}{3} - 1\frac{3}{5}\right) + \left(3\frac{2}{3} - 2\frac{2}{5}\right)$ **e)** $\left(\frac{17}{6} - 1\frac{2}{5}\right) - \left(\frac{3}{10} + \frac{1}{6}\right)$ **f)** $\frac{10}{3} - \left(1\frac{5}{6} + \frac{7}{3} + \frac{13}{6}\right)$

▶ Use a calculator to find:

a) $4\frac{5}{8} - 2\frac{9}{10}$ **b)** $5\frac{1}{8} + \frac{3}{4} - \frac{7}{5}$ **c)** $2\frac{2}{3} - \left(\frac{4}{5} + 1\frac{1}{9}\right)$ **d)** $\left(\frac{13}{12} + 1\frac{2}{3}\right) - \left(\frac{3}{4} + 1\frac{1}{2}\right)$

8 Jagruti wants to find $8\frac{1}{5} - 6\frac{2}{3}$ as a mixed number.

Her working is below. Her answer is correct.

$$8\frac{1}{5} - 6\frac{2}{3} = \frac{41}{5} - \frac{20}{3}$$

$$= \frac{123}{15} - \frac{100}{15}$$

$$= \frac{23}{15}$$

$$= 1\frac{8}{15}$$

a) Critique her method.

b) Show a more efficient method.

Thinking and working mathematically activity

Below are some calculations done by students. For each calculation:

- check to see if the answer is correct
- critique the method
- show a better method, and explain why your method is better.

①
$$17\frac{2}{9} - 2\frac{2}{3} = 17\frac{2}{9} - 2\frac{6}{9}$$
$$= \frac{155}{9} - \frac{24}{9}$$
$$= \frac{131}{9}$$
$$= 14\frac{5}{9}$$

②
$$2\frac{3}{4} - 4\frac{5}{8} + 3\frac{1}{6} = 2\frac{6}{8} - 4\frac{5}{8} + 3\frac{1}{6}$$
$$= -2\frac{1}{8} + 3\frac{1}{6}$$
$$= 3 - 2 + \frac{4}{24} - \frac{3}{24}$$
$$= 1\frac{1}{24}$$

③
$$3\frac{5}{6} - 2\frac{1}{4} + 1\frac{3}{8} = 3\frac{20}{24} - 2\frac{6}{24} + 1\frac{1}{8}$$
$$= 1\frac{14}{24} + 1\frac{1}{8}$$
$$= 1\frac{112}{192} + 1\frac{24}{192}$$
$$= 2\frac{136}{192}$$
$$= 2\frac{17}{24}$$

④
$$6\frac{3}{5} - \left(1\frac{1}{4} + 2\frac{9}{10}\right) = 6\frac{12}{20} - \left(1\frac{5}{20} + 2\frac{18}{20}\right)$$
$$= 6 - 3 + \frac{12}{20} - \frac{23}{20}$$
$$= 3 - \frac{11}{20}$$
$$= 2\frac{9}{20}$$

11.3 Multiplying and dividing fractions

Worked example 3

For each calculation, first state whether the answer will be larger or smaller than the first mixed number. Then find the answer. If it is larger than 1, write it as a mixed number.

a) $2\frac{2}{3} \times 1\frac{3}{4}$
b) $3\frac{1}{6} \div 1\frac{2}{3}$

a) The answer will be larger than $2\frac{2}{3}$	$1\frac{3}{4}$ is larger than 1. Multiplying a quantity by a number larger than 1 will give an answer larger than the quantity.

$$2\frac{2}{3} \times 1\frac{3}{4} = \frac{8^2}{3} \times \frac{7}{4_1}$$

$$= \frac{14}{3}$$

$$= 4\frac{2}{3}$$

Convert mixed numbers to improper fractions.

Cancel any factor that appears in both a numerator and a denominator.

If the answer is an improper fraction, you usually convert it to a mixed number.

b) The answer will be smaller than $3\frac{1}{6}$

$1\frac{2}{3}$ is larger than 1.

Dividing a quantity by a number larger than 1 gives an answer smaller than the quantity.

$$3\frac{1}{6} \div 1\frac{2}{3} = \frac{19}{6} \div \frac{5}{3}$$

Convert mixed numbers to improper fractions.

Change ÷ to × and invert the second fraction.

$$= \frac{19}{6^2} \times \frac{3^1}{5}$$

Cancel any factor that appears in both a numerator and a denominator.

$$= \frac{19}{10}$$

If the answer is an improper fraction, you usually convert it to a mixed number.

$$= 1\frac{9}{10}$$

Exercise 3

1 – 9

1 For each calculation, state whether the answer will be larger or smaller than the first number.

Then find the answer. If it is larger than 1, write it as a mixed number. Write fractions in their simplest form.

a) $3 \times \frac{2}{9}$ b) $\frac{2}{3} \times 8$ c) $11 \times \frac{1}{5}$ d) $\frac{3}{5} \times \frac{15}{3}$ e) $3\frac{2}{5} \times 3$ f) $\frac{16}{9} \times \frac{15}{8}$

g) $2\frac{1}{2} \times \frac{11}{9}$ h) $\frac{1}{9} \times \frac{3}{5}$ i) $\frac{4}{5} \times 1\frac{3}{4}$ j) $1\frac{2}{5} \times \frac{3}{7}$ k) $1\frac{3}{5} \times 1\frac{2}{3}$ l) $\frac{9}{5} \times 1\frac{2}{3}$

2 For each calculation, state whether the answer will be larger or smaller than the dividend (the first number). Then find the answer. If it is larger than 1, write it as a mixed number. Write fractions in their simplest form.

a) $\frac{1}{4} \div \frac{5}{8}$ b) $\frac{8}{11} \div 6$ c) $\frac{7}{8} \div \frac{3}{4}$ d) $\frac{3}{11} \div \frac{3}{8}$ e) $\frac{5}{6} \div 1\frac{3}{7}$ f) $\frac{11}{8} \div \frac{3}{4}$

g) $2\frac{3}{5} \div 3\frac{4}{7}$ h) $\frac{7}{5} \div \frac{5}{7}$ i) $1\frac{3}{10} \div \frac{2}{5}$ j) $2\frac{1}{8} \div \frac{7}{4}$ k) $\frac{11}{4} \div 1\frac{4}{11}$ l) $1\frac{2}{3} \div 1\frac{2}{9}$

3 Here are 5 cards: $1\frac{2}{3}$ $1\frac{1}{2}$ $\frac{3}{10}$ $1\frac{3}{5}$ $1\frac{1}{3}$

    a) Which fractions have a product of $\frac{1}{2}$?

    b) Which fractions have the highest product?

    c) Which fractions have the lowest product?

4 Find:

    a) $\left(2\frac{2}{3}\right)^2$ **b)** $\left(1\frac{1}{4}\right)^3$

> **Discuss**
>
> Explain why $\left(1\frac{1}{4}\right)^3$ does not equal $1\frac{1}{64}$.

5 Without doing the calculations, complete each statement using <, = or >.

    a) $\frac{2}{5} \times 2\frac{5}{6}\ \square\ 1\frac{1}{8} \times 2\frac{5}{6}$
    b) $\frac{7}{6} \times 2\frac{1}{2}\ \square\ \frac{8}{9} \times 2\frac{1}{2}$
    c) $2\frac{1}{4} \times \frac{5}{8}\ \square\ \frac{5}{8} \times 2\frac{1}{4}$

    d) $5\frac{1}{4} \div \frac{3}{10}\ \square\ \frac{3}{10} \div 5\frac{1}{4}$
    e) $2 \div \frac{3}{5}\ \square\ 1\frac{1}{2} \div \frac{3}{5}$
    f) $\frac{12}{11} \div \frac{11}{7}\ \square\ \frac{12}{11} \div 1\frac{4}{7}$

6 Look at this calculation. Explain the mistake and then correct it.

$$3\frac{3}{5} \div \frac{2}{9} = \frac{\overset{2}{\cancel{18}}}{5} \div \frac{2}{\underset{1}{\cancel{9}}}$$

$$= \frac{\overset{1}{\cancel{2}}}{5} \div \frac{1}{\underset{1}{\cancel{2}}}$$

$$= \frac{1}{5}$$

7 The perimeter of the rectangle is $12\frac{2}{5}$ cm. Work out the area.

Give your answer in its simplest form.

$3\frac{1}{2}$ cm

8 A tree was $\frac{5}{6}$ m tall when planted. This year the tree is $1\frac{1}{3}$ times taller.

How tall is the tree this year?

9 Find the answers. Write answers greater than 1 as mixed numbers.
Write fractions in their simplest form.

    a) $3\frac{3}{8} \div \left(\frac{2}{3} \times 1\frac{1}{4}\right)$
    b) $\frac{2}{5} \times \frac{7}{4} \div \frac{1}{2}$
    c) $2\frac{1}{3} \div \left(\frac{1}{3} + \frac{5}{6}\right)$
    d) $\frac{14}{11} + \frac{5}{8} \div 2\frac{3}{4}$

    e) $\frac{5}{6} + \frac{5}{3} \times \frac{8}{15}$
    f) $\left(1\frac{2}{5} + 1\frac{3}{10}\right) \times \frac{4}{3}$
    g) $2\frac{1}{2} \div \frac{3}{8} \times \frac{12}{5}$
    h) $\left(\frac{1}{3} - \frac{4}{15}\right) \div \left(\frac{13}{12} - \frac{1}{4}\right)$

10 Use a calculator to find:

    a) $\frac{4}{3} \div 4\frac{1}{9}$
    b) $3\frac{4}{5} \times 2\frac{3}{10}$
    c) $2\frac{1}{4} - \frac{1}{8} \times 1\frac{3}{5}$
    d) $2\frac{3}{7} \times \left(1\frac{1}{4} - \frac{4}{5}\right)$

Thinking and working mathematically activity

You can use partitioning to multiply two improper fractions. For example, you can write $2\frac{1}{2} \times 3\frac{3}{4}$ as $\left(2 + \frac{1}{2}\right) \times \left(3 + \frac{3}{4}\right)$. Show how to complete the calculation using this method.

You could write this using a multiplication grid.

Repeat the calculation the usual way, converting to improper fractions first.

Critique the two methods. Which method do you prefer, and why?

Would you always choose the same method? Can you find examples where each method is better?

Consolidation exercise

1 – 8

1. Sort the fractions below into fractions with terminating decimal equivalents and fractions with recurring decimal equivalents.

$$\frac{1}{45} \qquad \frac{6}{11} \qquad \frac{3}{80} \qquad \frac{17}{36} \qquad \frac{21}{28} \qquad \frac{7}{90} \qquad \frac{9}{27} \qquad \frac{8}{125}$$

2. Below are some denominators for fractions. For each denominator, write:

   • one fraction with a terminating decimal equivalent

   • one fraction with a recurring decimal equivalent.

   The fractions do not have to be in their simplest form.

   a) denominator 6   b) denominator 12   c) denominator 14

   d) denominator 26   e) denominator 30   f) denominator 45

3. Keir says, '15 has 3 as a prime factor, so any fraction with denominator 15 has a recurring decimal equivalent.'

   Is he correct? Explain your answer.

4. Maeve wants to answer the question, 'Does $\frac{19}{22}$ have a terminating or recurring decimal equivalent?'

   Her working is below, and her answer is correct.

   $$0.\ 8\ 6\ 3\ 6\ ...$$
   $$22\overline{\smash{\big)}19.\,^{19}0\,^{14}0\,^{8}0\,^{14}0}$$

   $\frac{19}{22}$ has a recurring decimal equivalent.

   Show how to answer the question using a quicker method.

5. Calculate the values.

   a) $1\frac{1}{9} \times 1\frac{1}{5}$ b) $1\frac{4}{5} \times 2\frac{1}{3}$ c) $3\frac{2}{15} - 1\frac{1}{3}$ d) $1\frac{13}{20} + 3\frac{3}{5}$ e) $2\frac{1}{8} \div 3\frac{2}{5}$

6. Jake has a ribbon that is $12\frac{1}{2}$ m long. He cuts it into smaller pieces. Each piece is $1\frac{2}{3}$ m long.

   a) Find how many pieces Jake can cut.

   b) Jake thinks there will be $\frac{1}{2}$ m of ribbon left over. Explain whether or not he is correct.

7 The fish in an aquarium are given $\frac{1}{8}$ kg of fish food each morning and $\frac{1}{5}$ kg of fish food in the afternoon.

A container has $4\frac{1}{3}$ kg of fish food. Find the number of whole days the fish food will last.

8 Marla and Takahiro pick strawberries. Marla puts $3\frac{1}{4}$ kg of strawberries in her basket but then eats $\frac{1}{5}$ kg of those.

Takahiro puts $1\frac{2}{5}$ kg into his basket, and then adds $\frac{3}{4}$ kg.

a) Write an expression for the difference between the masses of strawberries in Marla's and Takahiro's baskets.

b) Calculate the difference.

9 Pipe A is $2\frac{2}{5}$ m long has a mass of $4\frac{2}{3}$ kg. Pipe B is $2\frac{2}{9}$ m long and has a mass of $3\frac{4}{7}$ kg.

Compare the mass per metre for the two pipes – which is greater?

End of chapter reflection

You should know that...	You should be able to...	Such as...
A fraction in its simplest form is equivalent to: • a terminating decimal if its denominator's prime factors are only 2s and/or 5s • a recurring decimal if its denominator's prime factors include other numbers.	State, without calculating, whether a fraction has a terminating or recurring decimal equivalent.	Which of the fractions below have recurring decimal equivalents? Explain your reasoning. $\frac{7}{16}$ $\frac{9}{12}$ $\frac{15}{80}$ $\frac{4}{14}$ $\frac{6}{42}$
When a quantity is multiplied by a number larger than 1, the result is larger than the quantity. When a quantity is divided by a number larger than 1, the result is smaller than the quantity.	Before multiplying or dividing a quantity by a mixed number or improper fraction, state whether the answer will be larger or smaller than the quantity.	State whether each answer will be smaller or larger than the starting quantity. a) $5\frac{2}{3} \div 1\frac{1}{4}$ b) $\frac{2}{5} \times 2\frac{1}{2}$
Before multiplying or dividing with mixed numbers, convert them to improper fractions.	Multiply or divide mixed numbers and improper fractions.	Find: a) $2\frac{1}{5} \times 4\frac{3}{4}$ b) $1\frac{3}{8} \div \frac{22}{15}$
The usual order of operations applies to calculations with fractions.	Do calculations involving two or more types of operation, with mixed numbers and proper and improper fractions.	Find: a) $1\frac{5}{8} + \frac{7}{4} \div \frac{14}{15}$ b) $\left(\frac{12}{5} - 2\frac{1}{3}\right) \times 3\frac{1}{2}$

12 Probability 1

You will learn how to:

- Understand that the probability of multiple mutually exclusive events can be found by summation and all mutually exclusive events have a total probability of 1.
- Design and conduct chance experiments or simulations, using small and large numbers of trials. Calculate the expected frequency of occurrences and compare with observed outcomes.

Starting point

Do you remember...

- how to find the theoretical probability of a particular outcome happening? For example, a bag contains 3 red, 2 green and 5 orange stickers. What is the probability that a randomly-chosen sticker is red?
- how to work out the probability of an event not happening if you know the probability of it happening? For example, a bag contains 10 buttons, 4 of which are red. What is the probability that a randomly-chosen button is not red?
- how to make a tally chart? For example, what frequency does ||| || represent?
- how to solve linear equations? For example, find x if $3x - 4 = 5$
- how to add and subtract fractions, add and subtract decimals? For example, add $\frac{2}{5} + \frac{3}{10}$
- how to find equivalent fractions? For example, $\frac{3}{50} = \frac{?}{200}$
- how to convert between decimals, fractions and percentages? For example, write $\frac{21}{50}$ as a percentage.
- how to find relative frequency? For example, a dice is rolled 50 times and lands on a six 8 times. What is the relative frequency of obtaining a six?

This will also be helpful when...

- you learn more about probabilities of multiple events.

12.0 Getting started

Work in twos. Throw two six-sided dice and add the two numbers that you throw. Repeat 100 times. Make a tally chart of the answers which can be numbers from 2 to 12. Which answer is most common? Is this the same for the rest of your class?

$+$ $= 8$

12.1 Mutually exclusive events

When you throw a standard six-sided dice, it will land on one, and only one, number. It cannot land on, for example, both 3 and 5 at the same time. All the outcomes of throwing a dice (1, 2, 3, 4, 5 and 6) are said to be mutually exclusive. You could, however, get a number which is both an even number and a square number. These outcomes – throwing an even number and throwing a square number – are not mutually exclusive.

Key terms

Mutually exclusive outcomes of an event are a set of outcomes which cannot occur at the same time.

Any event can have more than one complete set of mutually exclusive outcomes. If a standard six-sided dice is thrown, a complete set of mutually exclusive outcomes could be throwing an odd number and throwing an even number. All six possible results are included. Another complete set could be throwing a number less than 3 and throwing a number greater than 2. Again, all six possible results are included.

The sum of the probabilities of a complete set of mutually exclusive outcomes is 1. This is because it is certain that one of the outcomes will happen, so if you add up the probabilities of all the mutually exclusive outcomes in a complete set then you must have certainty, which has a probability of 1.

Worked example 1

A coloured spinner has different sized sectors coloured green, red, yellow and blue and no other colours. The table shows the probability of the spinner stopping on a particular colour.

Colour	green	red	yellow	blue
Probability	0.3	0.4	x	$2x$

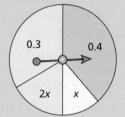

What is the value of x?

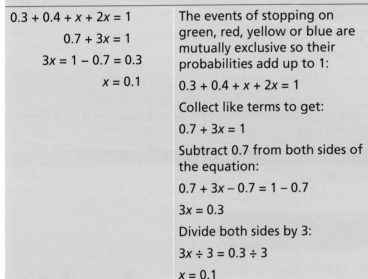

$0.3 + 0.4 + x + 2x = 1$

$0.7 + 3x = 1$

$3x = 1 - 0.7 = 0.3$

$x = 0.1$

The events of stopping on green, red, yellow or blue are mutually exclusive so their probabilities add up to 1:

$0.3 + 0.4 + x + 2x = 1$

Collect like terms to get:

$0.7 + 3x = 1$

Subtract 0.7 from both sides of the equation:

$0.7 + 3x - 0.7 = 1 - 0.7$

$3x = 0.3$

Divide both sides by 3:

$3x \div 3 = 0.3 \div 3$

$x = 0.1$

Exercise 1

1 A spinner has only blue, red and yellow sectors. The probability of the spinner landing on a particular colour is shown below. What is the value of x?

Colour	blue	red	yellow
Probability	$\frac{3}{10}$	$\frac{4}{10}$	x

2 A biased six-sided dice (with faces labelled 1, 2, 3, 4, 5 and 6) has the following probability of landing on each number:

Number	1	2	3	4	5	6
Probability	0.2	0.25	0.15	x	0.15	0.1

What is the value of x?

3 The train Jana takes to work on Monday will either be early, on time, late or cancelled. The table shows the probability of each of these events. What is the value of x?

	Early	On time	Late	Cancelled
Probability	x	0.65	x	0.05

4 A bag contains only red, yellow and green badges. A badge is chosen at random. The probability that the badge chosen is red is 0.22. The probability that the badge chosen is yellow is twice the probability that it is red.

a) What is the probability that the badge chosen is yellow?

b) What is the probability that the badge chosen is green?

c) What is the probability that the badge chosen is not green?

5 A biased six-sided dice (with faces labelled 1, 2, 3, 4, 5 and 6) is thrown. The probability of a 6 being thrown is $\frac{1}{12}$, the probability of throwing a 3 is the same as the probability of throwing a 4.

The probability of throwing a 2 is three times the probability of throwing a 6. The probability of throwing a 5 is twice the probability of throwing a 6. The probability of throwing a 1 is $\frac{1}{6}$.

a) What is the probability of throwing a 2?

b) What is the probability of throwing a 5?

c) What is the probability of throwing a number greater than 3?

d) What is the probability of not throwing a 3?

6 A sweet is chosen, at random, from a bag containing strawberry, orange and coffee chocolates and strawberry and orange jellies. The table shows the probabilities of choosing a particular type of sweet.

Sweet	Strawberry flavour	Chocolate	Orange flavour	Jelly	Coffee flavour
Probability	$\frac{7}{20}$	$\frac{11}{20}$	$\frac{11}{20}$	$\frac{9}{20}$	$\frac{2}{20}$

a) What is the sum of the probabilities in the table?

b) Is this a mutually exclusive set of outcomes? Give a reason for your answer.

c) There are 20 sweets in the bag, 2 of which are coffee flavour chocolates and 6 of which are orange jellies. How many strawberry flavoured chocolates are in the bag?

12.2 Experimental probability

Key terms

The **expected frequency** of an event = theoretical probability of the event occurring × number of trials.

For example, the probability of rolling a 1 on a fair six-sided dice is $\frac{1}{6}$. The dice is rolled 180 times.

The expected frequency of rolling a 1 is $\frac{1}{6}$ × 180 = 30

So you would expect a 1 to appear 30 times.

Worked example 2

Fausto is rolling a six-sided dice. He records the number of times it lands on a 6 after every 20 rolls and calculates the relative frequencies.

His results are shown in the table.

Frequency	4	16	22	27	32
Number of rolls	20	40	60	80	100
Relative frequency					

a) Calculate the relative frequencies after each 20 rolls and complete the table. Give your answers correct to two decimal places.

b) Plot the relative frequencies on a graph.

c) Fausto says the dice is biased. Do you agree? Give a reason for your answer.

d) Fausto rolls the dice a total of 600 times. If the dice was fair, find the expected frequency of it landing on a 6.

a) $4 \div 20 = 0.2$

$16 \div 40 = 0.4$

$22 \div 60 = 0.37$

$27 \div 80 = 0.34$

$32 \div 100 = 0.32$

Calculate relative frequency by dividing the frequency of rolling a 6 by the total number of rolls.

Frequency	4	16	22	27	32
Number of rolls	20	40	60	80	100
Relative frequency	0.2	0.4	0.37	0.34	0.32

b)

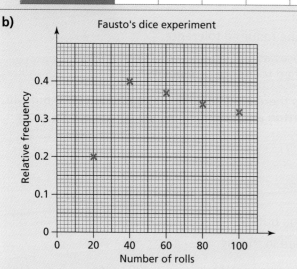

Each division on the relative frequency axis = 0.01

Each division on the number of rolls axis = 2

c) Fausto is correct.

The theoretical probability of the dice landing on a 6 is is $\frac{1}{6}$ or 0.167 (3 d.p.)

The relative frequencies of the observed values are tending towards 0.33, which indicates that the dice is biased.

Theoretical probability =

$$\frac{\text{number of favourable events}}{\text{total number of possible outcomes}}$$

d) Expected frequency = $\frac{1}{6} \times 600 = 100$ times

Expected frequency of an event = theoretical probability of the event occurring × number of trials.

The theoretical probability of a fair dice landing on a 6 is $\frac{1}{6}$ and the number of trials is 600. This gives an expected frequency of 100.

1 **a)** The probability that Leighton wins at darts is 0.45. He plays 40 games.
How many games would he be expected to win?

b) Maria rolls a fair six-sided dice 300 times. Calculate the expected frequency of it landing on 6.

c) The probability Barcelona FC win each game in La Liga is 0.7. How many games would they be expected to win in a season of 38 games?

d) The probability that a biased dice lands on a 3 is 0.1. Felicity rolls the dice 250 times.
Find the expected frequency of the dice landing on a 3.

2 A manufacturing company makes 100 000 bolts per week. The probability that each bolt is within a specified tolerance is 0.999.

a) Find how many bolts per week do **not** meet the required tolerance.

b) The company improves its manufacturing processes and only 50 bolts do not meet requirements each week. Find the new specified tolerance probability.

3 Li Wei wants to find out how many blue buttons there are in a large jar containing 1000 buttons. He performs an experiment where he takes a button at random, notes the colour and replaces it.

His results are shown in the table.

Number of blue button	28
Number of trails	120

a) Write the relative frequency of selecting a blue button.

b) Find an estimate of how many blue buttons are in the jar.

4 A spinner has a yellow sector, a blue sector and a green sector. It is spun 400 times.
The number of times it lands on yellow is recorded after every 100 spins.

The table shows the relative frequency of the spinner landing on yellow after each 100 spins.

Number of spins	100	200	300	400
Relative frequency of yellow	0.26	0.34	0.38	0.4

a) Find how many times the spinner landed on the yellow sector after 200 spins.

b) Copy and complete the graph.

c) Write down the best estimate of the theoretical probability of the spinner landing on yellow.

d) The spinner is spun a further 100 times and lands on yellow 50 times.

   i) Find the relative frequency of the spinner landing on yellow after 500 spins.

   ii) Plot this point on the graph.

   iii) What is the best estimate of the theoretical probability of the spinner landing on yellow after 500 spins?

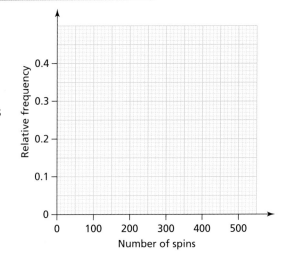

5 A four-sided dice is rolled 100 times. The number of times it lands on a 4 is recorded after every 20 rolls. The results are shown in the table.

Number of rolls	20	40	50	80	100
Number of 4s	5	14	18	27	30

a) Find the relative frequency of landing on 4 after 50 rolls.

b) Phillipe says the dice is a fair dice. Is Phillipe correct? Give a reason for your answer.

6 A spinner with red and white sectors is spun 150 times. The results are recorded after every 30 trials. The graph shows the relative frequency of the spinner landing on a white sector after each set of 30 trials.

a) The spinner landed on a white sector 15 times after 60 spins. Plot this point on a copy the graph.

b) How many times did the spinner land on a white sector after 120 spins?

c) Write down the best estimate from the graph of the probability of the spinner landing on a white sector.

d) Is the spinner a fair spinner? Explain your answer.

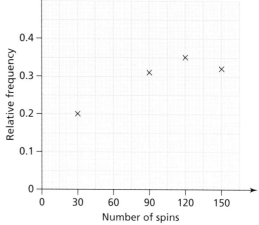

7 Joanna spins a spinner 50 times. She records the number of times her spinner lands on a green sector after every 10 trials and draws the graph below.

a) How many times did the spinner land on a green sector after 10 spins?

b) Find the best estimate from the graph of the probability of the spinner landing on a green sector.

c) How could Joanna improve her experiment to get an estimate of probability closer to the theoretical probability of the spinner landing on a green sector?

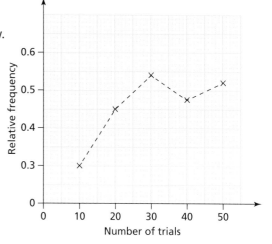

Thinking and working mathematically activity

Work individually or in pairs. Throw three dice and record whether at least one 6 is thrown. Repeat this trial 10 times. Find the relative frequency for the number of times at least one 6 is thrown.

Repeat for another 10 trials and find the relative frequency for all 20 trials.

Keep repeating the experiment until you have a total of 50 trials.

Plot your results on a graph.

Comment on the results. Draw a conclusion about how likely it is to get at least one 6 when three dice are thrown.

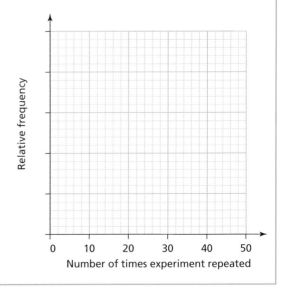

Consolidation exercise

1 **Vocabulary question** Copy and complete the sentences below with words from the box. Words may be used more than once.

total	event	probability	divided	trials	relative
	exclusive	expected	theoretical	multiplied	

a) The sum of all mutually .... outcomes of an .... is 1.

b) The .... frequency of a particular result of a trial is the number of times that result occurs .... by the .... number of trials.

c) You can use the .... frequency of an outcome as an estimator of the .... of that outcome.

d) The .... frequency of an event is the .... probability .... by the number of ....

2 In the morning, a school bus either arrives at school early, on time or late.

The probability that it arrives early is $\frac{1}{10}$ and the probability that it arrives

late is $\frac{7}{20}$. What is the probability that it arrives on time?

3 a) The probability that Xisco wins a game of chess is 0.4. He plays 65 games. How many games would he be expected to win?

b) Toni rolls a fair six-sided dice 576 times. Calculate the expected frequency of it landing on 4.

4 At Petticoat Junction, the rush hour train can arrive early, on time or late. The probability that it arrives late is twice the probability that it arrives early. The probability that it arrives on time is three times the probability that it arrives late. What is the probability that it arrives late? Give your answer as a fraction.

5 A spinner has three sectors: pink, orange and yellow. It is spun 200 times.
The number of times it lands on orange is recorded after every 50 spins.

The table shows the relative frequency of the spinner landing on orange after every 50 spins.

Number of spins	50	100	150	200
Relative frequency of orange	0.22	0.42	0.44	0.46

a) Find how many times the spinner landed on the orange sector after 150 spins.

b) Copy and complete the graph.

c) Write the best estimate of the theoretical probability of the spinner landing on orange after 200 spins.

d) The spinner is spun a further 50 times and lands on orange 33 times.

   i)   Find the relative frequency of the spinner landing on orange after 250 spins.

   ii) Plot this point on the graph.

   iii) Peter looks at the graph and says that half the spinner is coloured orange. Is Peter correct? Explain your reasoning.

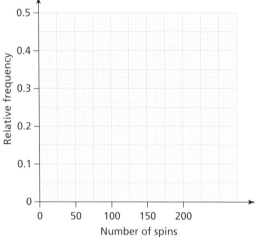

End of chapter reflection

You should know that...	You should be able to...	Such as...
The sum of the probabilities of a set of mutually exclusive outcomes of an event or trial is 1.	Calculate probabilities of an outcome if you know the probabilities of the other, mutually exclusive, outcomes.	A bag contains red, green and purple socks. A sock is chosen at random. The probability that the sock is red is $\frac{3}{10}$. The probability that the sock is green is $\frac{1}{2}$. What is the probability that the sock is purple?
The relative frequency of an outcome can be used as an estimate of its probability.	Calculate relative frequency as a fraction, decimal or percentage.	A bag contains red and yellow pens. A pen is taken from the bag, its colour noted and then it is replaced. This is repeated 100 times. The relative frequency of choosing a red pen is 38%. What is an estimate of the probability of choosing a yellow pen?
Expected frequency = theoretical probability × number of trials.	Calculate the expected frequency of an event.	How many 3's would you expect when a fair six-sided dice is rolled 120 times?
The relative frequency of an event tends to get closer to the theoretical probability as the number of trials increases.	Find the best estimate of theoretical probability from a relative frequency graph.	Draw a relative frequency graph to show how the relative frequency of a coin landing on heads changes as the number of trials increases. Is the coin biased or fair?

13

Equations and inequalities

You will learn how to:

- Understand that a situation can be represented either in words or as an equation. Move between the two representations and solve the equation (including those with an unknown in the denominator).
- Understand that a situation can be represented in words or as an inequality. Move between the two representations and solve linear inequalities.

Starting point

Do you remember…

- how to solve linear equations with unknowns on either or both sides?

  For example, solve $3x + 2 = 4x + 1$

- how to solve equations with the unknown in the numerator?

  For example, solve $\frac{2x}{4} = 6$

- how to simplify fractions?

  For example, simplify $\frac{15}{24}$

- how to use inequality symbols?

  For example, describe what $3 < x < 10$ means

- how to represent an inequality on a number line?

  For example, what is the inequality shown on this number line?

This will also be helpful when…

- you learn to manipulate and solve more complicated equations
- you learn to calculate bounds of accuracy
- you learn to plot graphs of inequalities or regions.

13.0 Getting started

Equations – sorted!

Play this game with a partner or in a group.

You each have a set of equations cards and an inequalities grid.

$x < -1$	$x = -1$	$-1 < x \leq 3$	$x > 3$
		$2x + 1 = 7$	

Place your cards face down on the table.

When the game starts, you turn over one card at a time from your set of cards.

You have three minutes to solve as many equations as you can and to place the cards in the correct box in the grid according to the solution.

For example, you pick a card that says $2x + 1 = 7$. You solve this to give $x = 3$, so you place this equation card in the $-1 < x \leq 3$ box in the grid.

At the end of the three minutes, compare your answers. Points are awarded as follows:

2 points for each equation correctly placed

−1 point for each equation incorrectly placed.

The winner is the player in your pair or group with the highest score.

13.1 Equations with fractions

Worked example 1

Solve the equations:

a) $\dfrac{24}{m} = 4$

b) $\dfrac{6}{x-2} = 2$

c) $7 = \dfrac{30}{4m+1} - 3$

a)

$$\dfrac{24}{m} = 4$$

$\times m$ $\times m$

$$\dfrac{24m}{m} = 4m$$

$$24 = 4m$$

$\div 4$ $\div 4$

$$6 = m$$

Multiply both sides of the equation by the denominator, m.

Remember that $\dfrac{m}{m} = 1$, so you can simplify $\dfrac{24m}{m}$ to 24.

Divide both sides by 4 to give the value of m.

Solution: $m = 6$

b)

$$\dfrac{6}{x-2} = 2$$

$\times (x-2)$ $\times (x-2)$

$$\dfrac{6(x-2)}{(x-2)} = 2(x-2)$$

$$6 = 2(x-2)$$

$\div 2$ $\div 2$

$$3 = x-2$$

$+2$ $+2$

$$5 = x$$

Multiply both sides of the equation by the denominator, $x - 2$.

Use brackets to make sure you multiply correctly.

Simplify $\dfrac{6(x-2)}{(x-2)} = 6 \times \dfrac{(x-2)}{(x-2)} = 6 \times 1 = 6$

Divide both sides by 2.

Add 2 to both sides.

Solution: $x = 5$

c)

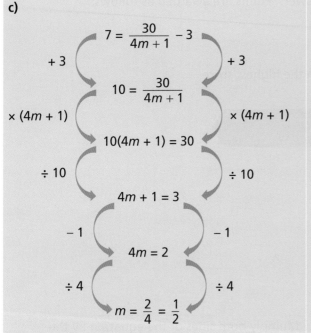

Add 3 to both sides to get a single fraction containing your unknown value on one side of the equation.

Multiply both sides by the denominator, $4m + 1$. Use brackets to make sure you multiply correctly.

Divide both sides by 10.

Subtract 1 from both sides.

Divide both sides by 4.

Solution: $m = \frac{1}{2}$

Exercise 1

1 Find the value of x in each of these equations.

a) $\frac{x}{2} = 10$ b) $\frac{10}{x} = 2$ c) $\frac{36}{x} = 9$ d) $\frac{36}{2x} = 6$

2 For each of these equations, say whether the answer given is true or false.

a) $\frac{24}{n} = 3$, $n = 8$ b) $\frac{12}{n + 1} = 2$, $n = 7$

c) $\frac{30}{n - 1} = 5$ $n = 7$ d) $\frac{100}{5n} = 5$, $n = 20$

3 Solve each of these equations to find the value of m.

a) $\frac{3}{m - 2} = 1$ b) $\frac{12}{m + 1} = 3$ c) $\frac{12}{m + 1} = -3$ d) $\frac{36}{2m + 3} = 4$

4 Selim and Sara are solving the equation $\frac{30}{2x - 5} = 2$.

Selim thinks the solution is $x = 10$.

Sara thinks the solution is $x = 8.75$.

Who is correct? Explain how you know.

5 Solve each of these equations to find the value of x.

a) $\frac{12}{x} + 4 = 7$ b) $\frac{9}{x + 1} + 2 = 3$ c) $\frac{7}{2x + 1} - 2 = 5$ d) $\frac{3}{x} + 2 = 4$

6 x students go to a photography club.

4 more students go to a gardening club than go to the photography club.

a) Write down an expression for the number of students who go to the gardening club.

A teacher shares 126 tulip bulbs equally between all the students in the gardening club. Each student receives 6 tulip bulbs.

b) Write an equation in x to represent this situation.

c) Solve your equation to find the number of students who attend the photography club.

7 Solve these equations.

a) $\dfrac{4}{x+3} + 2 = 0$ b) $\dfrac{6x}{x+1} = 3$ c) $\dfrac{3n}{n-2} = 4$ d) $\dfrac{4n}{2n+1} = 3$

8 Henri is trying to solve the equation $\dfrac{5x}{x+1} = 4$.

Write a set of instructions to tell him what he needs to do to solve this equation.

9 Nelson has y cups.

He has a bottle that contains 1200 millilitres of apple juice.

Nelson divides his apple juice equally between his cups.

He adds the apple juice in one cup to 200 milliltres of water to make himself a drink.

a) If he has 280 millilitres of drink, form an equation in y.

b) Solve your equation to find y.

> **Think about**
>
> Is it better to write non-integer solutions to equations as fractions or decimals?

Thinking and working mathematically activity

Working with a partner, make up a set of equations dominoes.

Remember that the question on one domino should match the answer on the next domino, and that the completed dominoes should form a loop.

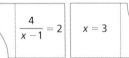

Make sure that:

- each of your equations has the unknown value in the denominator of the equation
- your equations all have different solutions, otherwise your dominoes will not work.

If you want to make your dominoes particularly challenging, use the same letter to represent the unknown value in each equation.

Test your dominoes by swapping them with another pair of students to see if they can solve your domino loop.

13.2 Solving inequalities

Key terms

When you solve an equation, you find specific values for which the equation is true. When you **solve an inequality**, you find a **range of values**, or **interval**, where the inequality is true for any value in that interval.

If you multiply or divide your inequality by a negative number, you must also **reverse the inequality sign** to keep the statement true.

For example, if $-x > 2$ then $x < -2$. Also, if $-k \leq -3$ then $k \geq 3$

Worked example 2

Solve each of these inequalities and show your answer on a number line:

a) $4x - 1 > 3$

b) $21 - 3x < 3$

c) $\dfrac{-x}{3} - 1 \leq 1$

d) $-5 < 2x + 1 \leq 5$

a)

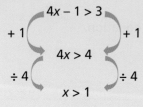

$4x - 1 > 3$

$+1 \qquad +1$

$4x > 4$

$\div 4 \qquad \div 4$

$x > 1$

Add 1 to both sides.

Divide both sides by 4.

$$\begin{array}{c} \circ \longrightarrow \\ \overset{-7\,-6\,-5\,-4\,-3\,-2\,-1\ 0\ 1\ 2\ 3\ 4\ 5\ 6\ 7}{\rule{0pt}{0pt}} \end{array}$$

b)

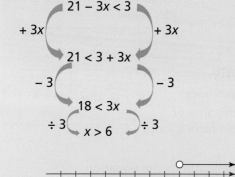

$21 - 3x < 3$

$+3x \qquad +3x$

$21 < 3 + 3x$

$-3 \qquad -3$

$18 < 3x$

$\div 3 \qquad \div 3$

$x > 6$

Add $3x$ to both sides.

Subtract 3 from both sides.

Divide both sides by 3.

$$\begin{array}{c} \circ \longrightarrow \\ \overset{-2\,-1\ 0\ 1\ 2\ 3\ 4\ 5\ 6\ 7\ 8\ 9}{\rule{0pt}{0pt}} \end{array}$$

c)

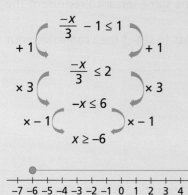

$\dfrac{-x}{3} - 1 \leq 1$

$+1 \qquad +1$

$\dfrac{-x}{3} \leq 2$

$\times 3 \qquad \times 3$

$-x \leq 6$

$\times -1 \qquad \times -1$

$x \geq -6$

Add 1 to both sides.

Multiply both sides by 3.

Multiply both sides by –1. You are multiplying by a negative number, so you need to reverse the inequality sign.

Use a closed circle to show that –6 is included in the solution. x is greater than or equal to –6.

$$\begin{array}{c} \bullet \longrightarrow \\ \overset{-7\,-6\,-5\,-4\,-3\,-2\,-1\ 0\ 1\ 2\ 3\ 4\ 5\ 6\ 7}{\rule{0pt}{0pt}} \end{array}$$

d)

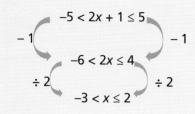

$$-5 < 2x + 1 \le 5$$

-1 $\quad$ -1

$$-6 < 2x \le 4$$

$\div 2$ $\quad$ $\div 2$

$$-3 < x \le 2$$

When there are three parts to an inequality, you use the same method as when there are two parts.

Subtract 1 from all parts of the inequality.

Divide all parts of the inequality by 2.

Use an open circle to show that −3 is not included in the solution. Use a closed circle to show that 2 is included in the solution. x is greater than −3 and less than or equal to 2.

From the answer you can see that x lies in a closed interval rather than an open interval.

Worked example 3

Solve $x + 7 > 3$ and $\frac{x}{2} \le 2$. Give your answer as a single inequality and show this on a number line.

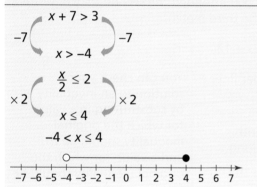

$$x + 7 > 3$$

-7 $\quad$ -7

$$x > -4$$

$$\frac{x}{2} \le 2$$

$\times 2$ $\quad$ $\times 2$

$$x \le 4$$

$$-4 < x \le 4$$

Subtract 7 from both sides.

Multiply both sides by 2.

As x can be both greater than −4 and less than or equal to 4, you can write the solution as a single inequality.

> **Did you know?**
>
> The symbols < and > were first introduced in mathematical books by Thomas Harriot. He was a mathematician who worked for Sir Walter Raleigh in the 16th century. Harriot adapted his symbols from previous mathematicians. The original symbols for less than and greater than were ⌐ and ⌐, respectively.

Exercise 2

1 Solve these inequalities and show your solutions on a number line:

    a) $3x < 21$ **b)** $\frac{x}{4} \ge -1$ **c)** $x - 5 < 6$ **d)** $8 - x < 3$

    e) $3x - 2 > 13$ **f)** $2x + 1 \le 15$ **g)** $13 - 2x \le 9$ **h)** $3 - 3x < -6$

2 Solve these inequalities:

    a) $2(x + 1) \le 10$ **b)** $3(x - 4) \ge 15$ **c)** $5(2x + 3) > -5$ **d)** $3(3x - 1) < 6$

3 Match each inequality with its correct solution interval.

$-2x > -8$	$x > -4$
$\dfrac{x}{4} - 2 \geq 1$	$x \geq -5$
$3x > -12$	$x \geq -2$
$5 - 3x \leq 20$	$x \geq 2$
$9 \geq 19 - 5x$	$x \geq 12$
$-7x - 3 \leq 11$	$x < 4$

4 A football club plays three matches. In total, less than 250 000 people attend the three matches. Exactly 95 000 attended the first match and less than 70 000 attended the second match. Write an inequality for the number of people, p, who attended the third match.

5 Show that $x = 8$ does not satisfy the inequality $5x - 2 < 3x + 12$.

6 Solve these inequalities. Show your solutions on a number line.

a) $3x - 5 \leq x + 1$ b) $7x + 1 < 3x - 7$ c) $7 - 2x \geq x - 5$

> **Tip**
>
> You can check your solution to an inequality by substituting in a value for which the original inequality should be true and a value for which the original inequality should be false.

7 Sasha's hair is 50 cm long. Her hair grows 1.5 cm per month. Sasha wants her hair to grow so that it is at least 78 cm long.

Sasha says she can write this as an inequality, where m is the number of months needed to grow her hair.

Which of these is the correct inequality? Explain your answer.

$50m + 1.5 \geq 78$	$50 + 1.5m \geq 78$	$50 + 1.5m > 78$	$50m + 1.5 < 78$

8 Solve these inequalities and show each of your solutions on a number line.

a) $4 < 2n < 10$ b) $-1 \leq y - 2 < 7$ c) $0 \leq \dfrac{x}{5} \leq 1$ d) $-3 < 2x - 1 \leq 7$

9 Marcus is solving the inequality $10 - 3y < 1$.

Here is his working.

$10 - 3y < 1$

$-3y < -9$

$y < 3$

Explain what mistake Marcus has made and show the correct solution.

10 Solve these inequalities:

a) $\dfrac{x}{2} + 2x \geq 5$

b) $25 + \dfrac{2x}{3} > 35 - x$

c) $\dfrac{3x}{4} + 5 \leq \dfrac{1}{2} - 6x$

Write an inequality using your number and x. You can choose the type of inequality.

Now try to construct another inequality for which your inequality is the solution.

For example, if you start with $x \leq 3$, you could write $2x \leq 6$, or $-x \geq -3$.

Now create inequalities with the same solution that have:

* an addition/subtraction
* two operations
* brackets
* a fraction
* an unknown on both sides
* three parts
  …. Combinations of the above.

Share your most complicated inequality with a partner and challenge them to solve it.

Consolidation exercise

1 Solve these equations.

a) $\dfrac{60}{x} = 12$ b) $\dfrac{72}{x + 7} = 6$ c) $\dfrac{9}{x - 5} = -3$ d) $\dfrac{12}{2x - 5} = -3$

2 I think of a number. I add 2 to my number. When I divide 42 by my result, I get the answer 6. What was the number I first thought of?

Explain how you worked it out.

3 Solve these equations.

a) $\dfrac{15}{x} - 3 = 2$ b) $\dfrac{45}{2x + 1} - 4 = -1$ c) $\dfrac{12}{2x + 1} + 8 = 4$ d) $\dfrac{x}{2x + 5} = 3$

4 Amy is n years old. Bella is 4 years older than Amy. When I divide Amy's age by Bella's age, I get an answer of 0.6.

a) Write an equation to show this problem.

b) Solve your equation to find Amy and Bella's ages.

5 Solve these inequalities. Show your solutions on a number line.

a) $3x > 18$ b) $5x - 4 \leq 6$ c) $17 < 3x + 2$ d) $6x + 7 > -5$

6 Are the solutions to these inequalities correct?
If they are not correct, work out the correct solution.

a) $0 < 9 + 3x$, $x < -3$ c) $\dfrac{x}{3} + 9 \geq 12$, $x \geq 9$

b) $2(5x - 4) > 27$, $x > 3.5$ d) $\dfrac{x}{4} - 3 \leq -7$, $x \leq -40$

7 I think of an integer, n.

30 minus twice my number is less than 15.

a) Write an inequality to show this problem.

b) Solve your inequality. What is the smallest possible value of my integer?

8 Solve each of these inequalities and show your solution on a number line.

a) $4x + 9 \geq 2x + 15$

c) $-1 \leq 2y + 1 < 11$

b) $-2 < \dfrac{x}{4} \leq 3$

d) $2x - \dfrac{1}{4} \leq \dfrac{3x}{4} + 1$

End of chapter reflection

You should know that...	You should be able to...	Such as...
If you have a fraction in an equation, multiplying the whole equation by the denominator of the fraction will make it easier to solve.	Solve equations involving fractions using the balance method.	Solve: $\dfrac{12}{3x - 4} = 6$
Inequalities can be solved using similar methods to solving equations and their solutions can be represented on number lines.	Solve inequalities using the same balance method used to solve equations, and use correct notation to show solutions on a number line.	Solve these inequalities and show your solutions on a number line: a) $4x + 1 \geq 9$ b) $-3 < 2x - 1 \leq 5$

14 Calculations

You will learn how to:

- Use knowledge of the laws of arithmetic, inverse operations, equivalence and order of operations (brackets and indices) to simplify calculations containing decimals and fractions.

Starting point

Do you remember…

- how to use the laws of arithmetic to simplify calculations?
  For example, find $\frac{4}{5} + \frac{3}{17} + \frac{1}{5}$ and $(17 \times 2.5) \times 4$
- how to follow the correct order of operations?
  For example, find $5 + 2^2 \times (\sqrt{4} - 1)$
- how to simplify calculations containing decimals or fractions?
  For example, use an efficient method to find $0.2 \times 5.9 \times 5$

This will also be helpful when…

- you use the laws of arithmetic, inverse operations, equivalence and order of operations with more complex calculations and algebra.

14.0 Getting started

The diagram shows a way of visualising the commutative law for addition.
It shows that $3 + 5 = 5 + 3$.

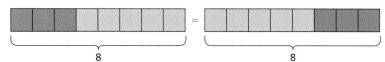

The table shows other laws of arithmetic and an example of each one.

For each law, create a diagram to show that the two sides are equal.
You could use either squared paper or isometric paper (for 3D drawing).

Law	Example
Commutative law for multiplication	$3 \times 5 = 5 \times 3$
Associative law for addition	$3 + (4 + 5) = (3 + 4) + 5$
Associative law for multiplication	$(4 \times 3) \times 2 = 4 \times (3 \times 2)$
Distributive law for multiplication	$4 \times (3 + 2) = 4 \times 3 + 4 \times 2$

Worked example 1

Use efficient methods to find:

a) $2.5 \times 32 \times 1.25$

b) $0.25 \times 7 + \frac{1}{4} \times 17$

c) $5 \times 3.7 \times \frac{4}{5} \div 4$

a) $2.5 \times 32 \times 1.25 = \frac{5}{2} \times 32 \times \frac{5}{4}$ $= 32 \times \frac{5}{2} \times \frac{5}{4}$ $= 32 \times \frac{25}{8}$ $= 25 \times 4$ $= 100$	You may find this calculation easier if you convert the decimals to fractions. Multiplication is commutative. You can change the order of the numbers to make the calculation simpler. Cancel a factor of 8. Then multiply.
b) $0.25 \times 7 + \frac{1}{4} \times 17 = \frac{1}{4} \times 7 + \frac{1}{4} \times 17$ $= \frac{1}{4} \times (7 + 17)$ $= \frac{1}{4} \times 24$ $= 6$	Notice that $0.25 = \frac{1}{4}$ Use the distributive law to rewrite using brackets. Follow the order of operations. Do the calculation in brackets first.
c) $5 \times 3.7 \times \frac{4}{5} \div 4 = 5 \times 3.7 \times 4 \div 5 \div 4$ $= 3.7 \times 5 \div 5 \times 4 \div 4$ $= 3.7$	Write $\frac{4}{5}$ as $\times 4 \div 5$ Multiplication and division have the same priority in the order of operations. You can change their order without changing the answer. $\times 5$ and $\div 5$ are inverse operations. $\times 4$ and $\div 4$ are also inverse operations. (Another way to write the calculation is: $3.7 \times \frac{4}{5} \times 5 \div 4 = 3.7 \times \frac{4}{5} \times \frac{5}{4} = 3.7$)

1 Using the correct order of operations, find:

a) $2.2 + 5 \times 10$

b) $6 \div \left(\frac{2}{3} - \frac{1}{6}\right)$

c) $\frac{2}{3} \times 5^2$

d) $(2.2 + 5) \times 10$

e) $6 \div \frac{2}{3} - \frac{1}{6}$

f) $\left(\frac{2}{5} \times 5\right)^2$

g) $\frac{1}{5} + \frac{3}{4} \times 0.8$

h) $\frac{9}{20} - (0.26 + 0.07)$

i) $\sqrt{\left(\frac{1}{4} + 0.75\right)} + 3.25$

2 Use efficient methods to find the answers. Show your working.

a) 0.2×165

b) $1 - \left(\frac{1}{4} + 0.4\right)$

c) $5^2 \times 0.84$

d) $0.75 \times 0.75 \times 32$

e) $\left(1\frac{7}{10} + 1.3\right)^3$

f) $1.25 \div \frac{5}{21}$

3 Use efficient methods to find the answers. Show your working.

a) $5.6 + (1.9 + 1.4)$

b) $\left(\frac{3}{11} + \frac{5}{8}\right) + \frac{8}{11} - \frac{5}{8}$

c) $6.03 - 4.9 - 2.1 + 2.97$

d) $\left(1\frac{7}{9} + \frac{5}{6}\right) + \left(2\frac{2}{9} - \frac{1}{6}\right)$

e) $12.28 + 2 \times 5.43 - 2.28$

f) $(1.92 + 0.25) + \frac{1}{4} - 1.92$

g) $0.44 + \frac{7}{10} - 0.34$

h) $6\frac{3}{100} - 1.9 - 2\frac{1}{10} - 1.03$

i) $11.03 - 11\frac{1}{20} - 2.03 + 2.05$

4 Use efficient methods to find the answers. Show your working.

a) $12 \times 8 \times 5 \times 0.25$

b) $\frac{3}{4} \times \left(7.2 \times \frac{4}{3}\right)$

c) $(1400 \times 0.23) \div 7$

d) $\frac{5}{6} \times \frac{1}{12} \div 5 \times 6$

e) $\frac{5}{3} \times 0.4 \times 18$

f) $3.5 \times 24 \times 2 \div 4$

g) $2.5 \div 3 \times 12 \times 27$

h) $1\frac{1}{3} \div 4 \times \frac{3}{4}$

i) $\left(\frac{2}{5} \times 26\right) \times 1.25$

5 Use the distributive law to find the answers. Show your working.

a) $2\frac{1}{4} \times 12$

b) $\frac{5}{12} \times 25 + \frac{5}{12} \times 11$

c) $1\frac{7}{12} \times 22 - \frac{1}{12} \times 22$

d) $\frac{1}{4} \times 1.1$

e) $2.15 \times 1.02 - 0.15 \times 1.02$

f) $9.9 \times \frac{3}{10}$

6 Group the calculations into pairs that give the same answer, and pairs that give different answers.

a)
$0.25 \times 7.8 \div 4$

$0.25 \div 4 \times 7.8$

b)
$48 \times 1\frac{3}{4}$

$48 \times 2 - 48 \times \frac{1}{4}$

c)
$4.5 \div 3 \times 1.2 \times 2$

$1.2 \times 2 \times 4.5 \div 3$

d)
$5.38 - 1\frac{8}{9} + 2\frac{1}{9} + 1.38$

$5.38 - 4 + 1.38$

e)
$\frac{5}{7} \div 5 \times 7$

$5 \div 5 \times 7 \div 7$

f)
$(2.11 \times 4.9) + (2.11 \times 5.1)$

4.22×10

7 Find the mistake in each calculation. Write the correct calculations.

a) $3.2 \times \left(\frac{3}{4} \div 4\right) = \left(3.2 \times \frac{3}{4}\right) \div 4$

$= 2.4 \div 4$

$= 0.6$

b) $1.5 + (0.2 \times 4) = (1.5 + 0.2) \times (1.5 + 4)$

$= 17 + 5.5$

$= 7.2$

> **Think about**
>
> Explain how the column method of multiplication uses the distributive law.
>
> Here is an example for you to think about:
>
> $$\begin{array}{r} 42 \\ \times\, 13 \\ \hline 126 \\ 420 \\ \hline 546 \end{array}$$

> **Discuss**
>
> Haruki, Iqbal and Jenna want to find $\frac{3}{4} \times 0.32$ without using a calculator.
>
> Haruki says, 'Convert $\frac{3}{4}$ to a decimal and then multiply.'
>
> Iqbal says, 'Convert 0.32 to a fraction and then multiply.'
>
> Try these two methods. Which do you prefer?
>
> Can you think of another method that is easier?

> **Thinking and working mathematically activity**
>
> Create questions that can be simplified by:
>
> - using laws of arithmetic
> - converting between fractions and decimals
> - using inverse operations.
>
> Include fractions and decimals in your questions.
>
> Swap questions with another student.

1 Niek buys three 0.4 kg bags of rice. He uses $\frac{3}{8}$ of each bag.

Find the total amount of rice he has left, in kilograms.

2 Kozue wants to find $\frac{4}{7} \times 42 \times \frac{2}{3}$. Her working is below.

$$\frac{4}{7} \times 42 \times \frac{2}{3} = \frac{4 \times 42 \times 2}{7 \times 3}$$

$$= \frac{336}{21}$$

$$= 16$$

Is her answer correct?

Show a more efficient way to do the calculation.

3 a) Which of the expressions below are equivalent to 2.25 × 48 × 1.25?

    A $48 \times 2\frac{1}{4} \times 1\frac{1}{4}$ **B** $100 \times \frac{5}{4}$ **C** $\frac{5}{4} \times \frac{9}{4} \times 48$

    D 2.25×60 **E** $\frac{48 \times 45}{8}$ **F** $\frac{5 \times 9 \times 48}{4 \times 4}$

  b) Use an efficient method to find 2.25 × 48 × 1.25

4 Gayatri has $\frac{3}{4}$ of a 900 g bag of flour and $\frac{3}{4}$ of a 1100 g bag of flour.

Use an efficient method to find the total amount of flour she has, in grams.

5 Use efficient methods to find the answers.

  a) $2\frac{5}{12} + 4\frac{3}{8} + \frac{7}{12} - 4\frac{3}{8}$ b) $5^3 + \sqrt{36^2} - 5^3$

  c) $2\frac{1}{2} \div 0.4 \times 20 \times 0.4$ d) $\frac{5}{8} \times 7.9 \times 24 \div 7.9$

6 In each pair below, only one calculation can easily be simplified.
Do this calculation without a calculator. Show your working.

Technology question Use a calculator to do the other calculation.

a)	4.16×8^2	4.16×5^2
b)	$\frac{5}{6} \times 0.4 \times 30$	$0.7 \times \frac{1}{6} \times 50$
c)	$0.75 \times 1.34 + \frac{5}{4} \times 1.34$	$\frac{2}{3} \times 8.7 + 1.3 \times 8.7$
d)	$4 \times \left(4.5 - \frac{4}{5}\right)^2$	$\left(2.6 - \frac{3}{5}\right)^2 \times 7$

End of chapter reflection

You should know that...	You should be able to...	Such as...
You can make some calculations easier by converting between equivalent fractions and decimals.	Convert between fractions and decimals where appropriate to make calculations easier.	Find: **a)** 145×0.2 **b)** $16 \times 0.75 \times 0.75$
You can make some calculations easier by combining inverse operations.	Combine inverse operations where appropriate to make calculations simpler.	Simplify and find: **a)** $\frac{5}{6} \times 6 \div 5$ **b)** $3.65 + \sqrt{17^2} - 3.65$
The associative laws says that: $(a + b) + c = a + (b + c)$ $(a \times b) \times c = a \times (b \times c)$ The commutative laws says that $a + b = b + a$ $a \times b = b \times a$ The distributive laws says that $(a + b) \times c = a \times c + b \times c$ $(a - b) \times c = a \times c - b \times c$	Use the laws of arithmetic (and the correct order of operations) to simplify calculations that include fractions, decimals, or both.	Simplify and find: **a)** $\frac{5}{8} + 1.74 + \frac{3}{8}$ **b)** $1.8 \times \left(5 \times \frac{17}{18}\right)$ **c)** $\left(1.18 \times \frac{3}{4}\right) + (0.25 \times 1.18)$

15

Pythagoras' theorem

You will learn how to:

- Know and use Pythagoras' theorem.

···

Starting point

- the types of triangle?

  For example, name the type of triangle that has a 90° angle.

- how to square a number, and how to square root a number?

  For example, find the value of 3^2, and use a calculator to find the value of $\sqrt{20}$

- you learn to solve geometric problems in three dimensions
- you learn to solve geometric problems requiring the use of both Pythagoras' theorem and the sine, cosine and tangent ratios.

15.0 Getting started

Look at the triangle drawn on square dotty paper.

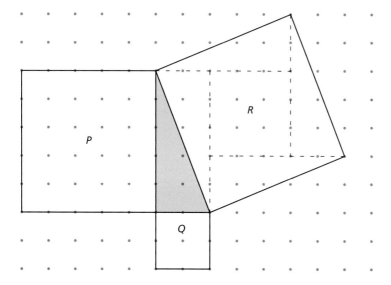

Three squares (*P*, *Q* and *R*) have been drawn on the edges of the sides of the triangle.

Square *R* has been divided into four triangles and a smaller square.

- Find the areas of squares *P*, *Q* and *R*.
- Draw your own right-angled triangles on square dotty paper. Add squares to each of the sides and find their areas.
- Can you find a relationship between the areas of the squares that works for all the right-angled triangles you have drawn?

Here is another right-angled triangle, with a square drawn on the third side.

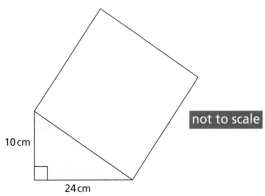

not to scale

10 cm

24 cm

Can you work out the length of the third side of the triangle?

Explain your method to a friend.

15.1 Pythagoras' theorem

Key terms

A **theorem** is a mathematical statement that has been proven.

The **hypotenuse** is the longest side in a right-angled triangle.

Pythagoras' theorem

For a right-angled triangle:

$$c^2 = a^2 + b^2$$

where c is the length of the hypotenuse and a and b are the lengths of the two shorter sides.

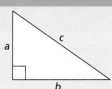

Worked example 1

a) Find the value of x in this right-angled triangle.

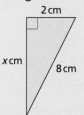

2 cm

x cm

8 cm

b) Use Pythagoras' theorem to find out whether or not this triangle is right-angled.

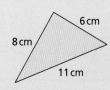

6 cm

8 cm

11 cm

a) $8^2 = 2^2 + x^2$ $64 = 4 + x^2$	The hypotenuse length is 8 cm. Write the relationship between the sides using Pythagoras' theorem.	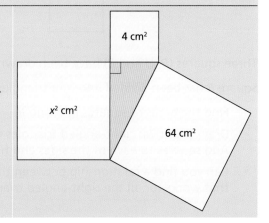
$x^2 = 64 - 4 = 60$ $x = \sqrt{60}$ $x = 7.7$ cm (1 d.p.)	Rearrange to make x the subject. If necessary, use a calculator to find the square root. Round the answer to one decimal place.	4 cm² x^2 cm² 64 cm²

b) $6^2 + 8^2 = 36 + 64$
$= 100$

$11^2 = 121$

100 does not equal 121, so the triangle is not right-angled.

If a triangle is right-angled, its sides obey Pythagoras' theorem: $a^2 + b^2 = c^2$, where c is the hypotenuse.

Find $a^2 + b^2$

Find c^2

Check to see if they are equal.

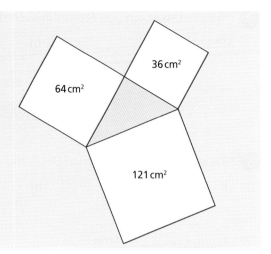

64 cm²   36 cm²   121 cm²

The diagrams in this exercise are not drawn to scale.

1 Find each unknown area, A cm².

a)

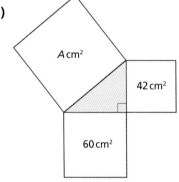

A cm²   42 cm²   60 cm²

b)

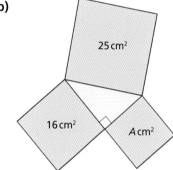

25 cm²   16 cm² A cm²

2 Find the length of the hypotenuse, h cm, in each right-angled triangle.

a)

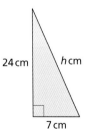

24 cm h cm   7 cm

b)

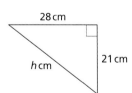

28 cm h cm   21 cm

c)

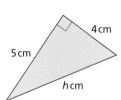

4 cm   5 cm h cm

3 Find the unknown side length, x cm, in each right-angled triangle.

a)

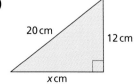

20 cm   12 cm x cm

b)

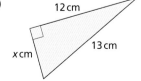

12 cm   13 cm x cm

c)

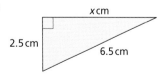

x cm   2.5 cm   6.5 cm

4 Find the unknown side length in each right-angled triangle.

a)

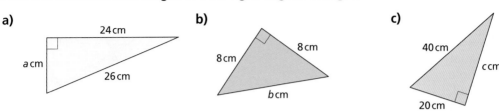

b)

c)

5 Find which of the three triangles is the odd one out. Explain your answer, showing any working.

a)

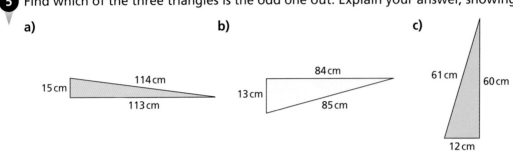

b)

c)

Thinking and working mathematically activity

Lucy says, 'If I multiply all three side lengths in this triangle by the same number, my new triangle will also be right-angled.'

Pedro says, 'If I add the same number to all three lengths, my new triangle will also be right-angled.'

Investigate their statements.

If you decide a statement is true, can you explain why?

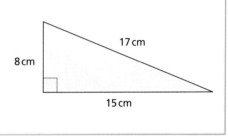

---

6 Katerina flies her kite, and it gets stuck at the top of a tree. The string is 18 m long, and when she pulls the string tight she can touch the end of the string to the ground 15 m from the tree.

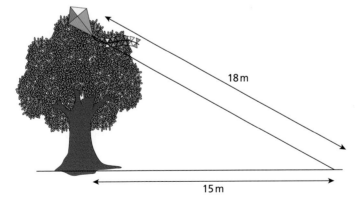

Find the height of the tree.

7 Find the length of a diagonal of this square.

1 m

8 Find the area of each triangle. In each part, you will need to use Pythagoras' theorem.

a)

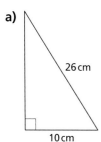

26 cm

10 cm

b)

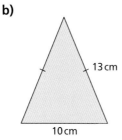

13 cm

10 cm

9 Without using a ruler, find the distance *PQ*.

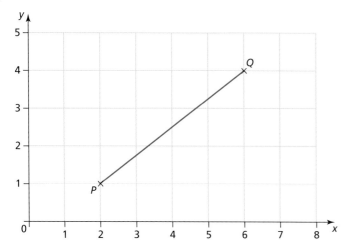

10 Here are four triangles.

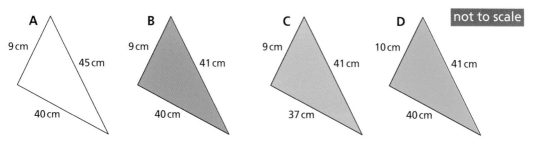

not to scale

A
9 cm
45 cm
40 cm

B
9 cm
41 cm
40 cm

C
9 cm
41 cm
37 cm

D
10 cm
41 cm
40 cm

Copy and complete the table by entering each triangle in the correct column.

Acute-angled	Right-angled	Obtuse-angled

Can you suggest a method that you could use to find out whether a triangle is obtuse-angled or acute-angled?

Consolidation exercise

The diagrams in this exercise are not drawn to scale.

1. The diagram shows a right-angled triangle.

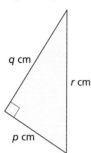

a) Choose the correct length of the hypotenuse.

   p cm   *q* cm   *r* cm

b) Choose the correct formula relating the lengths of the three sides.

   $p^2 = q^2 + r^2$ $q^2 = p^2 + r^2$ $r^2 = p^2 + q^2$

2. Write whether each statement is true or false. If a statement is false, write the statement correctly.

a) The hypotenuse of this triangle has length 100 cm.

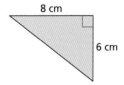

b) In this triangle, *x* is 13.

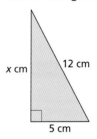

c) The length of side *s* of this triangle is given by $s = \sqrt{t^2 + u^2}$

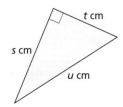

3 Find the side length x in each triangle.

a)

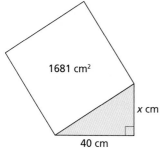

1681 cm²

x cm

40 cm

b)

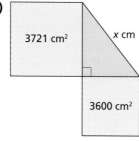

3721 cm²

x cm

3600 cm²

4 Explain why each of the triangles cannot exist.

a)

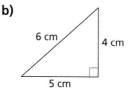

13 cm

5 cm

12 cm

b)

6 cm

4 cm

5 cm

5 Say whether each statement is always true, sometimes true, or never true, and explain your answer.

a) Pythagoras' theorem works for isosceles triangles.

b) Pythagoras' theorem works for equilateral triangles.

6 Jan and Naoko are checking whether a corner of their school sports field is right-angled. Starting from the corner, Jan walks 2 m along one side and Naoki walks 1.5 m along the other side. Find the straight line distance between the positions where they now stand.

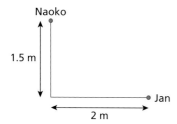

Naoko

1.5 m

Jan

2 m

7 Find the area of the parallelogram.

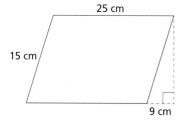

25 cm

15 cm

9 cm

End of chapter reflection

You should know that...	You should be able to...	Such as...
Pythagoras' theorem states that for a right-angled triangle, the square of the hypotenuse equals the sum of the squares of the other two sides.	Solve problems by finding an unknown side length of a right-angled triangle.	A ladder of length 2 m leans against a wall. The base of the ladder is 0.8 m from the wall. Calculate the height, h, of the top of the ladder.
	Use Pythagoras' theorem to find out whether or not a triangle is right-angled.	Find whether or not this triangle is right-angled.

16

Measures of averages and spread

You will learn how to:

- Use the mode, median, mean and range to compare two distributions, including grouped data.
- Interpret data, identifying patterns, trends and relationships, within and between data sets, to answer statistical questions. Make informal inferences and generalisations, identifying wrong or misleading information.

Starting point

Do you remember...

- how to calculate the mean, median, mode and range from a frequency table?

  For example, calculate the mean mark and the range for this data:

Mark	Frequency
7	5
8	9
9	5
10	1

- when to use each type of average?
  For example, which average would you use if you wanted to know which child has a height that is middle for the class?

- how to identify sets of discrete and continuous data?

  For example, are the marks shown in the frequency table above discrete or continuous data?

- how to complete a frequency table so that it has equal class intervals?

  For example, complete the classes in this frequency table so that they all have the same width.

Length, L cm	Frequency
$6 \leq L < 8$	
$8 \leq L < 10$	
___ $\leq L <$ ___	
___ $\leq L <$ ___	

This will also be helpful when...

- you learn how to find the median from a grouped frequency table using linear interpolation
- you learn how to find the mean from a grouped frequency table.

16.0 Getting started

Henri plays a game over and over again.

On each attempt, he gets a score between 1 and 20.

These are his scores for his first seven attempts:

Henri plays the game an eighth time.

What is his score if the mean of all eight scores is 11?

What could his score be if the range of all eight scores is 9?

What could his score be if the median of all eight scores is 10?

What is his score if the mean of all eight scores is 9.5?

What is his score if the median of all eight scores is 11?

16.1 Median, mode and range for grouped distributions

Key terms

When data is presented in a grouped frequency table, it is not possible to work out exact average values because the actual data values are not known.

- The **class interval** containing the median can be found. Remember that when there are n data values, the **median** value is in position $\frac{n+1}{2}$.
- The **modal class** can be identified – it is the one with the highest frequency.
- To find an **estimate of the range** from a grouped frequency table, subtract the smallest value from the first interval from the largest value in the last interval.

> ### Did you know?
>
> A value within a data set that is unusually small or large compared to the other data values is called an **outlier** (or an anomalous value).

Worked example 1

The frequency table shows the mass, m (grams), of some cakes at a cake sale.

a) Write down the interval that contains the cake with the median mass.

b) Write down the modal class interval.

c) Write down the largest possible value of the range.

Mass, m (grams)	Frequency
$300 \leq m < 400$	4
$400 \leq m < 500$	7
$500 \leq m < 600$	6
$600 \leq m < 700$	3

a) The median is in position $\frac{20+1}{2} = 10.5$ The median will be in the class interval $400 \leq m < 500$ because both the 10th and 11th data values are in this class interval.	When there are n data values, the median is in position $\frac{n+1}{2}$. Here, there are $4 + 7 + 6 + 3 = 20$ data values.
b) The modal class is $400 \leq m < 500$	This is the class interval that contains the largest frequency, which is 7 cakes.

c) The largest possible range would be

700 − 300 = 400 grams

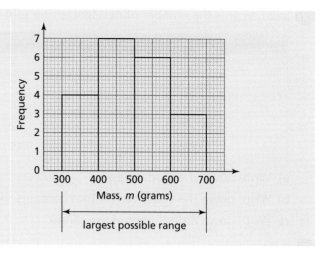

largest possible range

1 For each set of data, find which class interval contains the median value.

a)

Length, x (m)	Frequency
$2 < x \leq 3$	11
$3 < x \leq 4$	6
$4 < x \leq 5$	10
$5 < x \leq 6$	5

b)

Height, h (cm)	Frequency
$130 \leq h < 140$	3
$140 \leq h < 150$	11
$150 \leq h < 160$	4
$160 \leq h < 170$	2

c)

Number of books	Frequency
1 – 3	10
4 – 6	8
7 – 9	4
10 – 12	1
13 – 15	2

d)

Speed, x (km/h)	Frequency
$30 \leq x < 34$	7
$34 \leq x < 38$	10
$38 \leq x < 42$	15
$42 \leq x < 46$	5
$46 \leq x < 50$	3

2 Sven records the number of newspapers he sold on 15 days.

Number of newspapers sold	Frequency
81 – 85	4
86 – 90	5
91 – 95	4
96 – 100	2
TOTAL	**15**

a) Write down the modal class.

b) Write down the class interval that contains the median value.

c) Write down an estimate of the range.

3 Felix collects a sample of leaves and records their lengths in cm. Here are his results:

4.1, 2.8, 5.3, 8.6, 7.2, 3.5, 6.2, 3.4, 4.1, 1.8, 2.9, 5.7, 7.2, 4.6, 4.8, 3.1,
2.6, 4.4, 4.9, 5.1, 5.8, 6.3, 3.3, 3.0, 4.7, 5.1, 5.9, 6.3, 6.5, 5.6, 4.8, 4.1

a) Copy and complete the frequency table.

Length, l (cm)	Frequency
$1 \leq l < 2$	1

b) Write down the class interval that contains the median value.

c) Write down the modal class.

d) Compare the maximum range from the frequency table with the actual range.

4 The frequency diagram shows the mass of 80 objects.

a) Find the interval that contains the median mass.

b) Find the maximum possible range.

c) John says the median mass is 2.75 kg. Is he correct? Explain your answer.

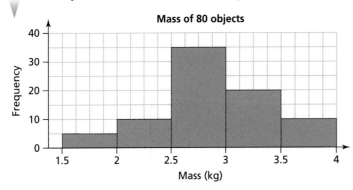

Worked example 1

Gray seals start to visit an island to have their pups. The number of pups born on the island are counted each year. In the first year, there are 6 pups. The number then increases by 95% each year.

a) Find the number of pups born on the island five years later. Write your answer to the nearest ten.

b) How many years after the start is the number of pups first higher than 2000?

a) $\quad$ $6 \times 1.95 = 11.7$ $\quad 11.7 \times 1.95 = 22.815$ $\quad 22.815 \times 1.95 = 44.4 \dots$ $\quad 44.4 \dots \times 1.95 = 86.7 \dots$ $\quad 86.7 \dots \times 1.95 = 169.1 \dots$ $\qquad\qquad \approx 170$ (to the nearest ten)	To find the number of pups after five years, increase the number by 95% five times. So, multiply 6 by 1.95 five times. Do not round numbers until the end.
or $6 \times 1.95^5 = 169.1 \dots$ $\qquad \approx 170$ (to the nearest ten)	Here is a quicker method for doing the calculation in one line.
b) After 6 years: $6 \times 1.95^6 = 329.8 \dots$ $\quad$ After 7 years: $6 \times 1.95^7 = 643.2 \dots$ $\quad$ After 8 years: $6 \times 1.95^8 = 1234.3 \dots$ $\quad$ After 9 years: $6 \times 1.95^9 = 2446.0 \dots$ The number of pups is first higher than 2000 after 9 years.	The answer must be more than 5 years, since there are only 170 pups after 5 years. Calculate the number of pups after 6 years, 7 years, and so on to find when the number is first higher than 2000.

Tip

Increasing by 95% is slightly less than doubling – so you might be able to see that after 6 or 7 years there will still be fewer than 2000. You could skip to calculating the number after 8 or 9 years.

Exercise 1

1 Write the multiplier for each calculation.

    a) increase by 18%   **b)** decrease by 30%   **c)** increase by 4%

    d) decrease by 95%   **e)** increase by 100%   **f)** increase by 355%

2 Write each answer to the nearest integer.

    a) increase 500 kg by 25%   **b)** increase 2100 m by 8%   **c)** decrease 366 minutes by 5%

    d) increase $840 by 120%   **e)** decrease 1620 ml by 16%   **f)** increase 40 km by 500%

3 A dress costs $40.00 in a shop. The price increases by 15%, and the new price later decreases by 15%.

   a) Find the final price of the dress.

   b) Write the absolute change in the price of the dress.

> **Discuss**
>
> A quantity increases by x%, where x is less than 100, and later decreases by x%. Is the result always, sometimes or never higher than the original quantity? Explain your answer.

4 a) An ash tree has a mass of 250 kg. Its mass increases by 5% each year.

   Find the mass of the ash tree, to the nearest kilogram, after:

      i) 1 year   ii) 5 years   iii) 10 years

   b) A maple tree has a mass of 250 kg. Its mass increases by 2% each year.

   Find the mass of the maple tree, to the nearest kilogram, after:

      i) 1 year   ii) 5 years   iii) 10 years

5 The value of a painting is $800. The value increases by 3% each year.

   a) Choose the formula for calculating the value of the painting after x years.

      $\$800 \times 0.03^x$ $\$800 \times 1.03^x$ $\$800 \times 3^x$

   b) Find the value of the painting after 8 years. Round your answer to the nearest $10.

6 A new car costs $15 000. The value of the car decreases by 15% each year.

   Calculate the value of the car at the ages below. Write each answer to the nearest hundred dollars.

   a) 1 year old   b) 2 years old   c) 6 years old

7 A new computer costs $600. Its value decreases by 30% each year.

   After how many years will the computer's value first drop below $200?

8 Describe a situation which each calculation below could represent.

   a) 50×1.1^4 b) 1200×0.95^6

9 **Technology question** There were no wild rabbits in Australia before 1859. In 1859, a farmer released 24 rabbits into the wild.

   After that, the population of rabbits in Australia increased by 21% each month.

   Create a spreadsheet showing the number of rabbits each month.

   a) Find how many **months** it took for the rabbit population to rise above:

      i) 100   ii) 1000   iii) 10 000

   b) Find the number of rabbits after 6 **years**. Write your answer to the nearest thousand.

   c) Eventually (after many years), the rabbit population stopped growing. Suggest one reason why.

> **Think about**
>
> A population of fruit flies doubles every hour. The population is 800 now. How long ago was the population 100?

Thinking and working mathematically activity

The bar represents a percentage increase. The original amount is represented by the dark green parts, and the new amount is the whole bar. Describe the percentage increase.

This bar can also represent a percentage decrease, where the whole bar represents the original amount and the darker parts represent the new amount. Describe the percentage decrease.

Use your answers to complete the sentence:

A ....% increase can be reversed by a ....% decrease.

Find the percentage change that reverses:

a 100% increase; a 300% increase; a 400% increase; a 20% decrease; a 90% decrease.

You could draw a bar for each example.

Consolidation exercise

1 Decide whether each statement is true or false. Explain your answers.

a) To increase a quantity by 150%, use the multiplier 1.5

b) To decrease a quantity by 20% four times, multiply the quantity by 0.2^4

c) To increase a quantity by 100% five times, multiply the quantity by 32

2 An antique chair is worth $450. Its value increases by 25% every 10 years.

Below are three suggested methods for calculating the value after 50 years.
Two of the methods give the correct answer.

Method A	Method B	Method C
Divide the quantity by 4 and add the result to the quantity. Repeat this five times.	Multiply the quantity by 0.25. Multiply the result by 5 and add this to the original quantity.	Multiply the quantity by 1.25. Repeat this five times.

a) State which method gives an incorrect answer.

b) Describe a more efficient method to find the value of the antique after 60 years, using a calculator.

Find the answer and round it to the nearest ten dollars.

3 On Wei and Jasmine's 10th birthdays, their parents start saving money for them.

Wei's parents save $100 on his 10th birthday.

On each birthday after that, they add $50 to the savings.

On the day after his 18th birthday, Wei receives all of the money saved.

Jasmine's parents save $100 on her 10th birthday.

On each birthday after that, they increase the amount of savings by 50%.

On the day after her 18th birthday, Jasmine receives all of the money saved.

a) After how many years does the amount saved by Wei's parents go above $400?

b) After how many years does the amount saved by Jasmine's parents go above $400?

c) Find how much money Wei receives after his 18th birthday.

d) To the nearest dollar, find how much money Jasmine receives after her 18th birthday.

4 When a giant pumpkin is growing, its mass increases by about 9% each day. One day, a farmer measures the mass of a pumpkin as 35 kg.

a) Find the mass of the pumpkin, to the nearest kilogram, 20 days later.

b) Find the mass of the pumpkin, to the nearest kilogram, 20 days *before* the farmer measured it.

5 Some quantities grow by a fixed percentage during equal lengths of time, while others grow by a fixed amount. For each quantity below, explain if it grows by a fixed percentage or a fixed amount.

a) the total paper thickness when you repeatedly fold a piece of paper in half (at equal time intervals)

b) the volume of water in a bucket being filled from a tap running at a constant rate

c) a population of bacteria (which reproduce by dividing in two) when there is plenty of space and food available to them

End of chapter reflection

You should know that...	You should be able to...	Such as...
A compound percentage is a repeated percentage change. You can find a compound percentage using a multiplier with a power. For example, the cost of an item increases by 5% each year for 6 years. Multiply the starting cost by 1.05^6 to find the final cost.	Use an efficient method to find the value of a quantity after a repeated percentage increase or decrease.	The population of a town is 12 000. The population decreases by 4% each year. Find the population after 7 years. After how many years will the population first be under 10 000?

18 Sequences

You will learn how to:

- Generate linear and quadratic sequences from numerical patterns and from a given term-to-term rule (any indices).
- Understand and describe nth term rules algebraically (in the form $an \pm b$, where a and b are positive or negative integers or fractions in the form $\frac{n}{a}$, n^2, n^3 or $n^2 \pm a$, where a is a whole number).

Starting point

Do you remember...

- how to generate terms of sequence using term-to-term rules?

  For example, find the 4th term of a sequence with first term – 2 and the term-to-term rule 'multiply by 2 and add 1'.

- how to generate sequences from spatial patterns?

  For example, draw the 5th term of this spatial pattern:

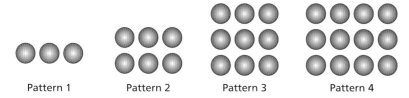

| Pattern 1 | Pattern 2 | Pattern 3 | Pattern 4 |

This will also be helpful when...

- you learn how to find the nth term of a quadratic sequence.

18.0 Getting started

The first few rows of Pascal's triangle are:

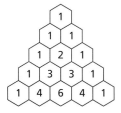

The first and last numbers in any row are 1.

You can find any number in the inside of the triangle by adding together the two numbers above it.

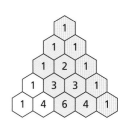

- Find the first eight rows of Pascal's triangle. Shade all the odd numbers in your triangle.
- What sequences do you notice along the diagonals of Pascal's triangle?
- Find the sum of the numbers in each row. Find the sum of the numbers in each row. What do you notice?

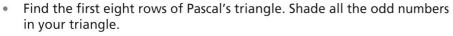

> **Did you know?**
>
> Pascal's triangle is named after French mathematician Blaise Pascal. However, mathematicians in other countries studied this triangle hundreds of years before Pascal. The same triangle is known as the Yang Hui triangle in China and the Khayyam triangle in Iran.

18.1 Generating sequences

Key terms

In a **linear sequence** the difference between successive terms is constant.

For example, 7, 11, 15, 19... is a linear sequence.
The first term is 7, and the **term-to-term rule** is 'add 4'.

In general, a linear sequence has a term-to-term rule of the form '+a' or '−a' where a is a fixed number.

The sequence 2, 5, 10, 17, 26 … is not linear.

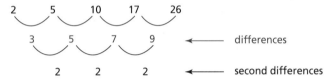

The **differences** are 3, 5, 7, 9, … The **second differences** are 2, 2, 2, …

The sequence 2, 5, 10, 17, 26 … is an example of a **quadratic sequence**.

In a quadratic sequence the second differences are constant.

You can use a term-to-term rule or the differences between terms to generate terms of a sequence.

Worked example 1

a) The first term of a sequence is 3. The term-to-term rule is 'square and subtract 1'. Find the third term of the sequence.

b) Write down the next two terms of the sequence 6, 8, 13, 21 …

c) Write down the next two terms of the sequence 20, 18, 15, 11 …

a) The third term is 63.	To get from one term to the next: first square and then subtract 1. $3^2 - 1 = 9 - 1 = 8$ $8^2 - 1 = 64 - 1 = 63$	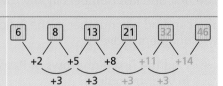
b) The next two terms are 32 and 46.	The differences between terms are 2, 5, 8 ….. The second differences are all 3. The next difference is $8 + 3 = 11$. The difference after that is $11 + 3 = 14$. The next number in the sequence is $21 + 11 = 32$. The number after that is $32 + 14 = 46$.	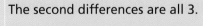 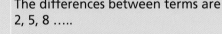 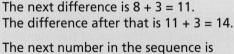 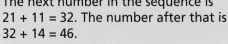

| c) The next two terms are 6 and 0. | The differences between terms are −2, −3, −4.

 The second differences are all −1.

 The next difference is −4 − 1 = −5. The difference after that is −5 − 1 = −6.

 The next number in the sequence is 11 − 5 = 6. The number after that is 6 − 6 = 0. | 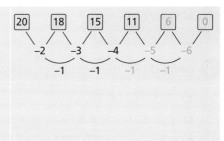 |

Worked example 2

Here are the first three patterns in a sequence.

a) Draw the 4th pattern in this sequence.

b) Explain why the number of squares in each pattern forms a quadratic sequence.

c) Find the number of squares in Pattern 6.

Pattern 1   Pattern 2   Pattern 3

| a)
 Pattern 4 | Look at the patterns. What is the same? What is different? | Pattern 4 will have a 4-by-4 square in the middle and one extra square at each corner. |

| b) The number of squares forms the sequence 5, 8, 13, 20 …

 The differences between terms are 3, 5, 7 …

 The sequence is quadratic because the second differences are constant (they are all 2). | Find the differences between the terms in the sequence.

 Then find the second differences.

 In a quadratic sequence, the second differences are constant. |  |

| c) The 5th term will be 20 + 9 = 29

 The 6th term will be 29 + 11 = 40 | The next difference is 7 + 2 = 9. The difference after that is 9 + 2 = 11.

 The next number in the sequence is 20 + 9 = 29. The number after that is 29 + 11 = 40. | 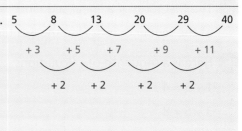 |

1 Write down the next three terms in each sequence.

a) 2, 5, 10, 17, 26, …

b) 1, 10, 23, 40, 61, …

c) 0.5, 1, 2.5, 5, 8.5, …

d) 0.3, 2.3, 6.3, 12.3, 20.3, …

e) −3.5, −2, 2.5, 10, 20.5, …

f) −4, −5, −7, −10, −14, …

g) 40, 37, 32, 25, …

h) 0.8, 0.65, 0.4, 0.05, …

2 Which of these sequences is the odd one out? Explain why.

A 1, 4, 9, 16, 25, …

B 1, 4, 7, 10, 13, …

C 1.6, 2, 2.8, 4, 5.6, …

D 1.6, 2, 3.4, 5.8, 9.2, …

3 Here are some patterns in a sequence made from squares.

a) Draw Pattern 5.

b) How many squares are there in Pattern 6?

c) Decide whether or not the number of squares in each pattern forms a quadratic sequence. Give a reason for your answer.

4 The first term of a linear sequence is 2.5. The 6th term of the sequence is 4.5.

...2.5........ ....4.5......

Archie thinks that the second term is 2.9. Leah thinks that the second term is 3.
Who is correct? Explain why.

> **Hint**
>
> What is the difference between 2.5 and 4.5?
>
> What should the value of '?' be?
>
>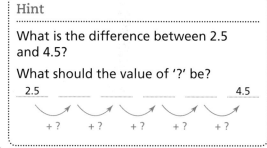

5 The first term of a linear sequence is $\frac{1}{10}$. The 4th term of the sequence is $\frac{5}{2}$. Find the 5th term.

6 Here is a sequence of calculations:

1 + 3 = 4

3 + 5 = 8

5 + 7 = 12

7 + 9 = 16

a) Write down the next two calculations in this sequence.

b) Write down one comment about the numbers on the right-hand side of these calculations.

c) Copy and complete: ………….. + ………….. = 54

7 A sequence of calculations begins:

$1 = 1$

$1 + 3 = 4$

$1 + 3 + 5 = 9$

$1 + 3 + 5 + 7 = 16$

a) Write down what you notice about the answers in this sequence.

b) Write down the next calculation.

c) Could one of the calculations in this sequence have an answer of 289?
Give a reason for your answer.

8 The first term of a sequence is –2. The term-to-term rule is 'square and subtract 6'.
Comment on the terms in this sequence.

9 The first term of a sequence is –2. The term-to-term rule is 'square and subtract 3'.

a) Find the first six terms of this sequence.

b) Write down what you notice about the terms of this sequence.

10 The first term of a sequence is 3. The term-to-term rule is 'square and subtract 5'.
Find the 4th term.

11 The first term of a sequence is 2. The term-to-term rule is 'square and subtract 5'.
Hope says that the 4th term of this sequence is –21. Is she correct? Explain your reasoning.

Thinking and working mathematically activity

- The diagram shows the first three patterns in a sequence.

Pattern 1   Pattern 2   Pattern 3

1   3   6

Draw Patterns 4 and 5.

- These patterns produce the sequence of triangular numbers: 1, 3, 6, 10, 15, …
Find the first ten triangular numbers.

- Find the sum of pairs of consecutive triangular numbers. For example, $3 + 6 = 9$
What is special about all these sums?

- Find a way for finding any triangular number directly from its position number.
Use your method to find the 100th triangular number.

Discuss
....

The term-to-term rule for a sequence is 'add 1 and square'.
The first term is 1. Is this a quadratic sequence?

18.2 nth term rules

Key terms

A **linear sequence** has an **nth term rule** of the form $an \pm b$, where a and b are positive or negative integers or fractions.

For example, the sequence with the nth term rule $4n + 3$ is a linear sequence.

A **quadratic sequence** has n^2 and no higher powers in its nth term rule.

For example, the sequence with the nth term rule $n^2 + 2$ is a quadratic sequence.

A sequence that is based on cube numbers is called a **cubic sequence**.

You can use the nth term rule of a sequence to find terms in that sequence.

You can also use the terms of a sequence to find the nth term rule of that sequence.

Worked example 3

a) A rule for generating the nth term of a sequence is $n^2 + 1$. Find the 4th term of the sequence.

b) A sequence has nth term rule n^3. Find the 5th term of the sequence.

c) Find an expression for the nth term of the sequence that begins 4, 7, 12, 19, ...

a) $4^2 + 1$ $= 16 + 1$ $= 17$	To find the 4th term: substitute $n = 4$ in the nth term rule.	
b) $5^3 = 125$	To find the 5th term: substitute $n = 5$ in the nth term rule.	
c) The nth term rule is $n^2 + 3$	The difference between the terms are 3, 5, 7.... The second differences are constant (2, 2, ...). The sequence is quadratic so compare the terms with the square numbers.	Square numbers: 1 $\xrightarrow{+3}$ 4 $\xrightarrow{+5}$ 9 $\xrightarrow{+7}$ 16 $\quad n^2$ $+3 \downarrow \quad +3 \downarrow \quad +3 \downarrow \quad +3 \downarrow$ Sequence: 4 $\xrightarrow{+3}$ 7 $\xrightarrow{+5}$ 12 $\xrightarrow{+7}$ 19 $\quad n^2 + 3$

Exercise 2

1 Match each nth term rule to the sequence it describes.

$-2n + 5$	1, 3, 5, 7, 9, ...
$n^2 + 2$	3, 1, −1, −3, −5, ...
n^2	1, 4, 9, 16, 25, ...
n^3	1, 8, 27, 64, 125, ...
$2n - 1$	3, 6, 11, 18, 27, ...

2 Find the difference between the 3rd and 5th terms of the sequence with nth term rule n^3.

3 The nth term rule of a sequence is $\frac{n}{7}$. Helen says that the sum of the first four terms is $\frac{4}{7}$

  a) Explain why Helen is incorrect.

  b) Calculate the correct sum of the four terms.

4 Terms in sequence A have the nth term rule $45 - 2n$.

  Terms in sequence B have the nth term rule $n^2 - 17$.

  Which term in sequence A is equal to the 6th term in sequence B?

5 Georgia is finding the nth term rule of the sequence $-4, -1, 4, 11, \ldots$

  Georgia compares the sequence with the square numbers to find the nth term rule.

  Use Georgia's method to find the nth term rule of this sequence.

6 Here are some quadratic sequences. By comparing the terms with the square numbers, find the nth term rule of each sequence.

  a) 3, 6, 11, 18, 27, …   **b)** 0, 3, 8, 15, 24, …   **c)** 6, 9, 14, 21, …   **d)** −9, −6, −1, …

7 The nth term rule of a sequence is n^3. Which of these numbers are not terms in the sequence?

1	6	8	9	12	27	64

8 Sort the numbers below in the correct cell in the Carroll diagram.

8	4	64	12	15

	In sequence with nth term rule n^2	Not in sequence with nth term rule n^2
In sequence with nth term rule n^3		
Not in sequence with nth term rule n^3		

9 The nth term rule of a sequence is $\frac{n}{8}$.

  Find the sum of the 11th term and the 13th term of this sequence.

10 Decide whether each statement is true or false. Correct the mistakes in any false statements.

  a) In the sequence $\frac{1}{7}, \frac{2}{7}, \frac{3}{7}, \ldots$ the nth term rule is $\frac{n}{7}$.

  b) In the sequence $\frac{1}{14}, \frac{2}{14}, \frac{3}{14}, \ldots$ the nth term rule is $\frac{2n}{7}$.

  c) In the sequence with nth term rule $\frac{n}{10}, \ldots$ the ratio 5th term : 10th term = 1 : 2.

> **Think about**
>
> Does the nth term rule $n^2 + 4$ give a different sequence to the rule $(n + 4)^2$?

Thinking and working mathematically activity

- Find an example of a quadratic sequence where the second term is an odd number.
- Find a number that appears in the sequence with nth term rule n^3 and the sequence with nth term rule $n^2 + 2$.
- Find an example of a quadratic sequence where the first four terms are negative.
- Can the first two terms of a quadratic sequence be even numbers? Give a reason for your answer.
- Can 0 be a term in a quadratic sequence? Give a reason for your answer.
- Can 1 be the first term in a quadratic sequence if the nth term rule is $n^2 + a$ where a is a positive integer? Give a reason for your answer.

Consolidation exercise

 1 For each sequence, decide if the terms are always positive, always negative or sometimes positive and sometimes negative.

a) First term = 2
Term-to-term rule is 'square and subtract 5'

b) The nth term rule is n^3

c) The nth term rule is $n^2 + 4$

d) First term = −4
Term-to-term rule is 'multiply by 2 and subtract 1'

2 Match the terms to the sequence that generates them.

a) The nth term rule is n^3	**v)** The 5th term = 1 The 10th term = 2
b) The nth term rule is $n^2 + 4$	**w)** The 2nd term = 8 The 5th term = 29
c) The nth term rule is $\frac{n}{5}$	**x)** The 1st term = 1 The 10th term = 46
d) The nth term rule is $5n - 4$	**y)** The 3rd term = 8 The 5th term = 24
e) The nth term rule is $n^2 - 1$	**z)** The 2nd term = 8 The 4th term = 64

3 Here is a sequence of patterns made from grey squares.

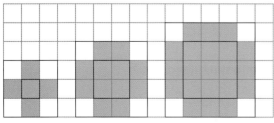

Pattern 1   Pattern 2   Pattern 3

a) Copy and complete the table.

Pattern	1	2	3	4	5
Number of grey squares					

b) Raya says that Pattern 7 has 100 squares. Is she correct?
Show how you worked out your answer.

4 **a)** Write down the first five terms of the sequence $n^2 + 8$.

b) Jo says that the term-to-term rule is 'add 3'. Show that Jo is incorrect.

5 The second term of a linear sequence is 17. The 6th term of the sequence is 29.

Which of these numbers is a term in the sequence?

21   14   34   1

6 Look at the nth term rules for the four sequences here:

n^3 $9n + 8$ $n^2 + 4$ $\dfrac{n}{2}$

True or false?

125 is a term in all of these sequences. Explain your answer.

End of chapter reflection

You should know that...	You should be able to...	Such as...
A sequence can be generated by continuing a pattern or by using a term-to-term rule.	Use term-to-term rules and differences between terms to continue sequences.	Find the next term of the sequence that begins 16, 14, 10, 4, …
A linear sequence can be described by a nth term rule of the form $an \pm b$, where a and b are positive or negative integers or fractions.	Use some given values in a linear sequence to find: • other terms in the sequence • the term-to-term rule • an expression for the nth term rule.	The second term in a linear sequence is 40. The 5th term is 16. Find: **a)** the first term of the sequence **b)** an expression for the nth term rule.
Some sequences have nth term rules of the form $\dfrac{n}{a}$, n^2, n^3 or $n^2 \pm a$.	Use nth term rules to find the value of specific terms.	Find the 10th term in the sequences generated by these nth term rules: **a)** $n^2 + 2$ **b)** n^3
	Find the nth term rule of quadratic, cubic and other sequences.	Find the nth term rule of these sequences **a)** $\dfrac{1}{5}, \dfrac{2}{5}, \dfrac{3}{5}, \dfrac{4}{5}$ ….. **b)** 2, 5, 10, 17….

19

Area and measures

You will learn how to:

- Know and use the formulae for the area and circumference of a circle.
- Estimate and calculate areas of compound 2D shapes made from rectangles, triangles and circles.
- Know and recognise very small or very large units of length, capacity and mass.

Starting point

Do you remember…

- how to use the formula $C = \pi d$ to find the circumference of a circle?

  For example, find the circumference of a circle with radius 10 cm.

- how to find the area of parallelograms, trapeziums and compound shapes made from them?

  For example, find the area of this shape.

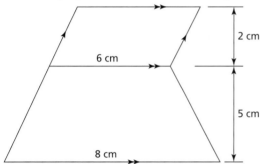

- how to convert between metric units of length?

  For example, convert 2.9 km to centimetres.

- how to convert between ordinary numbers and numbers in standard form?

  For example, write 0.00047 in standard form.

This will also be helpful when…

- you find the area and volume of more complex 3D shapes.

19.0 Getting started

How can we find the area of a circle?

- Look at the circle of radius r divided into eight sectors.

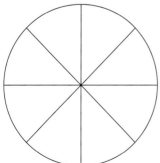

Separate the sectors and rearrange as shown.

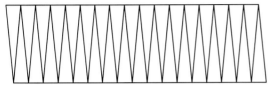

The sectors very roughly form a rectangle.

- If you divided the circle into 32 sectors, the arrangement would be a closer approximation to a rectangle.

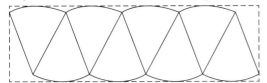

If you divided the circle into thousands of sectors it would be much closer to a rectangle.

- Think about this rectangle made from all the sectors - what are its dimensions?
- The formula for the area of a circle is $A = \pi r^2$. Can you see how this relates to the area of the rectangle?

19.1 Area and circumference of a circle

Key terms

The length around a circle is called the **circumference.**

Circumference $= \pi d$ or $2\pi r$.

Area of circle $= \pi r^2$.

When solving circle problems, use the π button on your calculator to give a more accurate answer for the circumference and area of a circle.

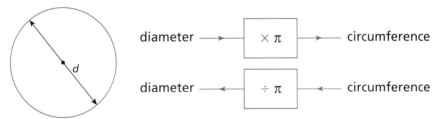

diameter $\longrightarrow$ $\times \pi$ $\longrightarrow$ circumference

diameter $\longleftarrow$ $\div \pi$ $\longleftarrow$ circumference

Circumference $= \pi d$
Circumference $= 2\pi r$

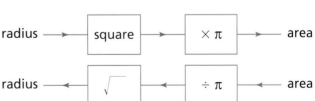

radius $\longrightarrow$ square $\longrightarrow$ $\times \pi$ $\longrightarrow$ area

radius $\longleftarrow$ $\sqrt{}$ $\longleftarrow$ $\div \pi$ $\longleftarrow$ area

Area $= \pi r^2$

Worked example 1

a) Find the area of:

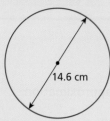

b) Work out the area and perimeter of:

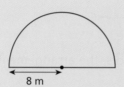

a) $r = 14.6 \div 2$ $r = 7.3$ cm $A = \pi r^2$ $= \pi \times 7.3^2$ $= 167.4$ cm^2 (1 d.p.)	Divide the diameter by 2 to work out the radius. Write down the formula for the area and substitute the value for r.	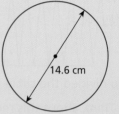
b) Area of full circle is $\pi r^2 = \pi \times 8^2$ $= 201.06\ldots$ m^2 So, the area of the semicircle is $\dfrac{201.06}{2} = 100.5$ m^2 (1 d.p.) Circumference of full circle is $\pi d = \pi \times 2 \times 8$ $= 50.265\ldots$ m So, the perimeter of the semicircle is $\dfrac{50.265\ldots}{2} + 2 \times 8$ $= 41.1$ m (1 d.p)	The area of a semicircle is half the area of a full circle. Write down the formula, then substitute in the value for r. The perimeter of a semicircle is half the circumference of the full circle plus the diameter of the circle.	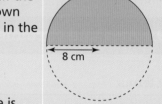

Worked example 2

A circle has an area of 28 cm^2. Find the radius.

$A = \pi r^2$ $28 = \pi \times r^2$ $r^2 = \dfrac{28}{\pi}$ $r^2 = 8.912676813$ $r = \sqrt{8.912676813}$ $r = 3.0$ cm (1 d.p.)	Write down the formula for the area and substitute the value for A. Divide by π to make r^2 the subject. Take the square root to find the value of r.	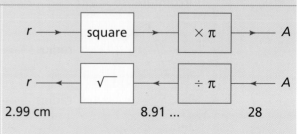

Did you know?

About 4000 years ago people knew that the value of π was close to $3\frac{1}{7}$ or $3\frac{1}{8}$.

About 2200 years ago, Archimedes discovered that π is less than $3\frac{10}{71}$ but greater than $3\frac{1}{7}$

About 1500 years ago, a Chinese mathematician, Zu Chongzhi, calculated π to be $\frac{355}{113}$.

About 300 years ago, a Japanese mathematician, Seki Takakazu, calculated π to be less than 3.141459265359.

About 30 years ago, using a super computer the value of π was calculated to 10 million places.

Today we can calculate the value of π to trillions of decimal places!

Exercise 1

The diagrams in this exercise are not to scale.

1 Find, correct to one decimal place, the area of a circle with:

  a) radius = 5.2 cm　　　　**b)** radius = 3.7 m　　　　**c)** radius = 26 mm

  d) diameter = 12.8 cm　　**e)** diameter = 7.6 m　　　**f)** diameter = 86 mm

2 Copy and complete the table.

Radius	Diameter	Area (rounded to nearest whole number)
6.8 cm		
	11.6 cm	
2.1 m		
	134 mm	

3 Priya has completed her homework on circumference and area of circles.

Are her answers correct?

If an answer is wrong, then explain what mistake she has made.

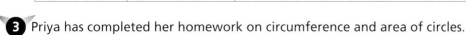

4.6 cm

$C = \pi d$
$C = \pi \times 4.6$
$\quad = 14.5$ cm

$A = \pi r^2$
$A = \pi \times 4.6^2$
$\quad = 66.5$ cm

12.4 cm

$C = \pi d$
$C = \pi \times 12.4$
$\quad = 39.0$ cm

$A = \pi r^2$
$A = \pi \times 6.2^2$
$\quad = \pi \times 12.4$
$\quad = 39.0$ cm

4 Find the difference in the areas of these two shapes.

Give your answer correct to one decimal place.

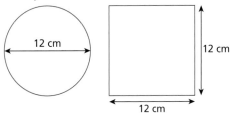

For each of the shapes shown below, find:

i) the perimeter of the shape   **ii)** the area of the shape.

Give your answers correct to one decimal place.

a)
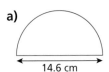

b)

c)
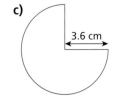

6 The circumference of a circle is 100 m.

a) Calculate its diameter to 1 decimal place.

b) Calculate its area to 3 significant figures.

Isabella makes a necklace piece by using two circles.

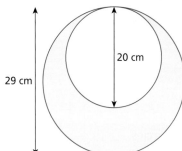

She paints a part of it yellow, as shown.

Calculate the area she paints. Give your answer to 1 decimal place.

8 Calculate the diameter of a circle with an area of 280 cm².

Give your answer to one decimal place.

9 This is a picture of two unicycles.

The first unicycle wheel has a radius of 48 cm.

The second unicycle wheel has a radius of 33 cm.

How much further does the first unicycle go than the second one when their wheels rotate once?

Give your answer to the nearest centimetre.

10 It takes 28 g of flour to make dough for a pizza with a diameter of 100 mm.

How much flour do you need to make the same pizza, but with a diameter of 200 mm?

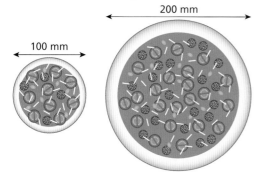

200 mm

100 mm

11 Calculate the circumference of a circle with an area of:

a) 120 cm²   **b)** 314 mm²   **c)** 478 m²

Give your answers to 1 decimal place.

> Tip
>
> Calculate the radius first.

12 There are two green areas in the park. They have equal areas.

Tom runs around the circular path B. What distance does he run?

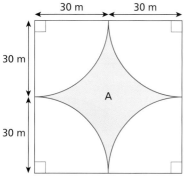

30 m   30 m

30 m

A

30 m

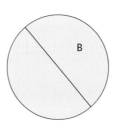

B

To calculate the area and the perimeter of the shaded region, Joanne does the following calculations. Do you agree with her? Explain why.

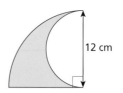

12 cm

Area of shaded part	Perimeter of shaded part
Area of 12 cm $= \dfrac{\pi \times 12^2}{4} = 36 \times \pi = 113.0973355$ cm² Area of semicircle $\bigg($ 12 cm $= \dfrac{\pi \times 6^2}{2}$ $= 18 \times \pi = 56.54866776$ cm² Shaded area $= 113.0973355 - 56.54866776$ $\qquad\qquad = 56.5$ cm² (1 d.p.)	Perimeter of 12 cm $= \dfrac{2 \times \pi \times 12}{4} = 6 \times \pi = 18.84955592$ cm Perimeter of semicircle $\bigg($ 12 cm $= \dfrac{\pi \times 12}{4}$ $= 3 \times \pi = 9.4247777961$ cm Perimeter of shaded part $\qquad = 18.84955592 - 9.4247777961$ $\qquad = 9.4$ cm (1 d.p.)

Think about

Ben is running along the blue track. Samira is running along the red track. Who runs the furthest?

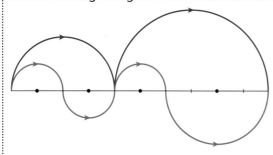

Design a similar track which will give the same result.

 Thinking and working mathematically activity

This shape is made up of semicircles.

If the radius of the small semicircle is 1 cm, find the area of the shaded shape. Leave your answer as a multiple of π.

Investigate the link between the radius of the small semicircle and the shaded area.

What would the radius of the small semicircle be if the shaded area is 400π cm²?

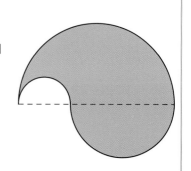

Worked example 3

This shape is made from two semicircles and a trapezium.

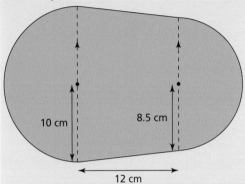

Calculate the area, giving your answer to the nearest whole number.

Area of small semicircle:

Area $= \frac{1}{2}\pi r^2$

$= \frac{1}{2} \times \pi \times 8.5^2$

$= 113.49$ cm^2

Area of large semicircle:

Area $= \frac{1}{2} \times \pi \times 10^2$

$= 157.10$ cm^2

Divide the shape into semicircles and a trapezium.

For each semicircle, write down the formula and substitute the value of the radius.

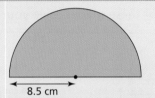

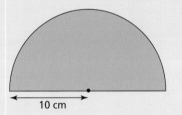

Area of trapezium:

Area $= \frac{1}{2}(a + b)h$

$= \frac{1}{2} \times (17 + 20) \times 12$

$= 222$ cm^2

The length of the parallel sides of the trapezium are $2 \times 8.5 = 17$ cm and $2 \times 10 = 20$ cm.

Write down the formula for area of a trapezium and substitute.

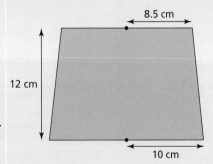

Total area $= 113.49 + 157.10 + 222$

$= 492.59$

$= 493$ cm^2 (to nearest whole number)

Add together the three areas. Write the final answer to the nearest whole number.

The diagrams in this exercise are not to scale.

1　Calculate the area that is shaded in these diagrams.

a)

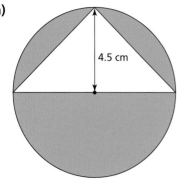

4.5 cm

b)

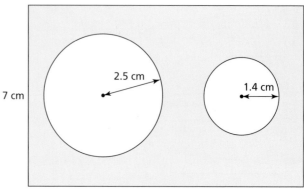

7 cm

2.5 cm

1.4 cm

11 cm

2　Find the area of these shapes. Give your answers to 1 decimal place.

a)

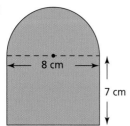

8 cm

7 cm

b)

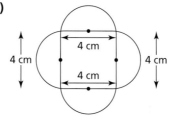

4 cm

4 cm

4 cm

4 cm

4 cm

3　The diagram shows a shape formed from a triangle and a semicircle.

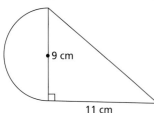

9 cm

11 cm

Mae said the area of this shape is 81 cm².

Is she correct? Explain your answer.

4　The diagram shows a shape formed from a rectangle and two quarter circles.

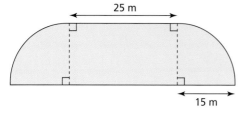

25 m

15 m

Calculate the area of the shape.

5 Find the percentage of the circle that is shaded.

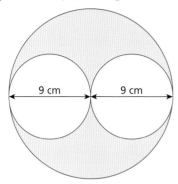

9 cm   9 cm

6 Find the percentage of the rectangle that is shaded.

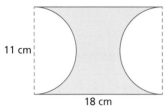

11 cm

18 cm

Give your answer to 1 decimal place.

7 A new lawn is created in a city centre as shown. Zac plans to cover the area with grass seed.

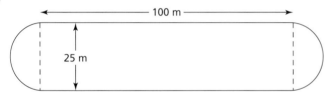

100 m

25 m

It is recommended that you use 1 kg of grass seed for every 8 m².

Zac says he will need 400 kg of grass seed.

Is Zac correct? Show how you worked out your answer.

A rectangular sheet of metal 140 cm by 75 cm has 3 cm radius discs stamped out of it.

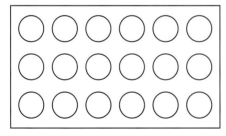

a) What is the largest number of discs that can be stamped out of the rectangular sheet?

b) What area of metal would be wasted? Give your answer to 1 decimal place.

Thinking and working mathematically activity

A running track needs to:
- have a perimeter of 400 metres
- have two equal straight sides and two equal semi-circular sides.

Design a possible running track.

Find the area it encloses.

Investigate the area enclosed by other running tracks.

Investigate whether it is possible for the track to enclose an area of 13 000 m².

19.3 Small and large units

Key terms

A **tonne** is a unit used for measuring mass.

A **light year** is a unit used for measuring large distances (such as distances between planets).

A **byte** is a unit used in computing (for example in measuring the size of a file).

Small units
milli (m) .... is 1 thousandth (10^{-3})
micro (μ) .... is 1 millionth (10^{-6})
nano (n) .... is 1 billionth (10^{-9})
Examples
micrometre (μm)   nanometre (nm)
millisecond (ms)   nanosecond (ns)
microlitre (μl)

Large units
1 light year (ly) = 9 460 730 472 580 800 m
1 tonne (t) = 1000 kg
mega (M).... is 1 million (10^6)
giga (G).... is 1000 million = 1 billion (10^9)
tera (T)..... is 1 million million (10^{12})
Examples
kilotonne (kt)   megatonne (Mt)
megabyte (MB)   gigabyte (GB)   terabyte (TB)

Converting between small units:

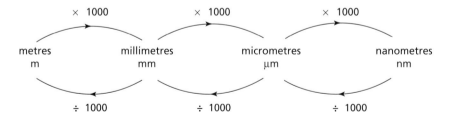

Converting between large units:

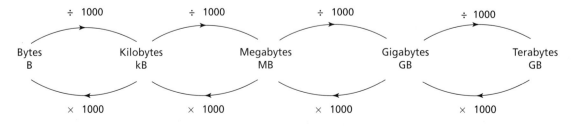

Discuss

Think of some objects whose mass might be recorded in tonnes.

Did you know?

The star Alpha Centauri A is approximately four light years away from Earth.

Worked example 4

a) Write 0.4 megatonnes in tonnes.

b) An atom has diameter 0.1 nm. A human hair has a width that it 700 000 times greater.
   Find the width of the hair in millimetres.

a) 0.4 Mt = 0.4 × 1000 = 400 kt = 400 × 1000 = 400 000 t	Multiply by 1000 to convert between megatonnes and kilotonnes. Multiply by 1000 again to convert between kilotonnes and tonnes.	
b) Width of human hair is 0.1 × 700 000 = 70 000 nm 70 000 nm = 70 000 ÷ 1000 = 70 mm = 70 ÷ 1000 = 0.07 mm	Multiply 0.1 nm by 700 000 to find the width of the hair in nanometres. Divide by 1000 to convert this to micrometres. Divide by 1000 again to convert to millimetres.	

Exercise 3

1 Match each item to the most suitable units.

The mass of a grain of salt	ly
The diameter of an atom	μm
The amount of water in a drop	mg
Distance from Earth to a star	μg
The mass of an atom	Mt
Thickness of a microorganism	μl
The mass of Earth's moon	nm
The amount of milk in a teaspoon	MB
The size of a computer file	ml

2 Put each of these lists into order of size, starting with the smallest.

    a) 8 mm, 8 nm, 8 ly, 8 km, 8 μm,

    b) 5 t, 5 μg, 5 kg, 5 mg

    c) 64 GB, 64 TB, 64 MB

    d) 48 ms, 0.2 s, 2400 μs, 1.7 ms

3 Complete the gaps in these conversions.

    a) 16 tonnes = ... kg   **b)** 3 MB = ... B

    c) 5 milliseconds = ...seconds   **d)** 8 μg = ... g

    e) 0.000016 m = ... μm   **f)** 75 μm = ... nm

    g) 5 100 000 t = ... Mt   **h)** 0.8 Mt = ... t

    i) 6 TB = ... GB   **j)** 0.00085 g = ... μg

4 Milk contains 122 mg of calcium per 100 ml. How many grams of calcium are in a 2 litre carton of milk?

5 A computer memory card holds 128 MB of data.

Write 128 MB in Bytes. Give your answer in standard form.

6 The thickness of a piece of paper is 230 μm.

    a) Write this thickness in metres. Give your answer in standard form.

    b) Find, in metres, the thickness of a pile of 5000 pieces of this type of paper.

7 It is important for us to have Vitamin A in our diet.

Men need 0.7 mg of Vitamin A per day. Women need 0.6 mg per day.

Carrots contain 8.35 μg of Vitamin A per gram.

Ollie is cooking carrots for himself and his wife. He says, 'If I cook 170 g of carrots for our dinner, we will have sufficient vitamin A for both of us.'

Is Ollie correct? Explain your answer.

8 Ahmed's computer takes 1 second to download 14.5 MB of data.

Find how long it will take his computer to download a film that is 1.3 GB.

9 50 μl of water drips into a bowl every second.

The bowl has a capacity of 1 litre.

Find how long it will take for the bowl to fill. Give your answer in hours.

10 The recommended daily dose of Vitamin K is 1 μg for each kilogram of body weight.

Half a cup of cooked spinach contains 444 μg Vitamin K.

Half a cup of raw spinach contains 72 μg Vitamin K.

Lauren has mass 28 kg. She eats one quarter of a cup of cooked spinach.

Seamus has mass 46 kg. He eats one quarter of a cup of raw spinach.

Have both Lauren and Seamus eaten the recommended daily dose of Vitamin K? Give reasons for your answer.

- Research the recommended daily doses of Vitamin A, B, C, D, E for men and women and what foods contain these vitamins.
- Think about your own diet, do you have the recommended dose of vitamins?
- Create a meal menu that will contain sufficient vitamins for a family of four.

Consolidation exercise

1 Calculate the area of each circle, giving your answer to 1 decimal place.

a)

5.8 cm

b)

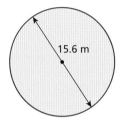

15.6 m

c)

1.1 mm

2 Decide which shape has the larger area: a circle of radius 6 cm or a square of side 10 cm. Explain your answer.

3 a) Change 260 micrograms to grams.

   b) Change 0.084 tonnes to kilograms.

   c) Change 0.4 GB to kilobytes. Give your answer in standard form.

4 To calculate the radius of a circle with area of 100 cm², Matilde does the following calculations:

Area of circle = πr^2

$100 = \pi \times r^2$

$r = \sqrt{100} \div \pi$

$r = 10 \div \pi = 3.2$ cm (1 d.p.)

Is Matilde correct? Explain your answer.

5 A circular pond has a radius of 8 m. Its base is to be covered with a layer of concrete. 60 kg of concrete mixture is needed to cover 1 m². The concrete mixture is supplied in 50 kg bags. How many bags are needed?

6 Calculate the area of these shapes. Give all answers to 1 decimal place.

a)

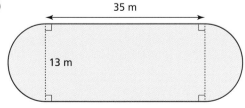

35 m

13 m

b)

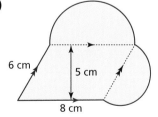

6 cm

5 cm

8 cm

c)

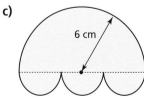

6 cm

7 A grain of rice has a mass of 29 mg. How many grains of rice are in a 1 kg bag of rice?

8 Ben says that 3 μm = 3×10^{-3} mm.

Is Ben correct? Explain your answer.

End of chapter reflection

You should know that...	You should be able to...	Such as...
Area of a circle = πr^2 Circumference of a circle = πd	Calculate the area and circumference of a circle.	Calculate the area and circumference of a circle with radius 8 cm.
Compound shapes can be made from triangles, rectangles and circles.	Calculate the area of a compound shape by splitting it into simpler shapes.	Calculate the area of this compound shape. 25 cm 8 cm
Milli, micro and nano units can be used to measure very small quantities. Kilo, mega and giga units can be used to measure very large quantities.	Use very small and very large units of length, capacity and mass.	Convert 5 cm to nanometres.

20

Presenting and interpreting data 2

You will learn how to:

- Record, organise and represent categorical, discrete and continuous data. Choose and explain which representation to use in a given situation:
  - dual and compound bar charts
  - pie charts
  - infographics
  - scatter graphs.
- Interpret data, identifying patterns, trends and relationships, within and between data sets, to answer statistical questions. Make informal inferences and generalisations, identifying wrong or misleading information.

Starting point

Do you remember...

- how to draw and interpret dual and compound bar charts?

  For example, how many gold medals did female athletes get?

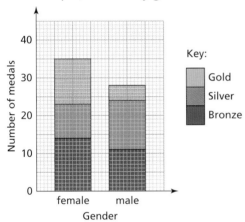

- how to interpret an infographic?

  For example, the infographic shows the ingredients required to make a San Sebastian cheesecake for two people. How much sugar would you need to make a cheesecake for five people?

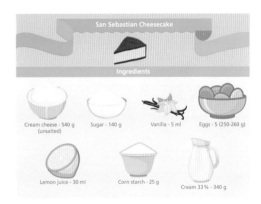

- how to draw and interpret a pie chart?

  For example, draw a pie chart to show the favourite types of soft drink for 50 people.

Drink	Frequency
Orange juice	10
Cola	16
Lemonade	20
Blackcurrent	4

- how to draw and interpret scatter graphs?

For example, describe the relationship between mass and wing area.

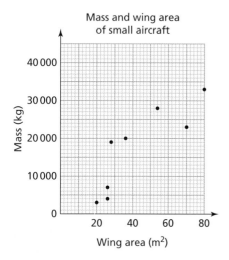

Mass and wing area of small aircraft

- you convert a bar graph or pictogram into a pie chart
- you use and interpret sophisticated infographics.

20.0 Getting started

Florence Nightingale was a British nurse who helped treat wounded soldiers in the Crimean War (1853–1856). Her use of statistics helped to save the lives of many soldiers.

Here is her graph (known as a rose diagram) that compares the number and causes of death in two 12-month periods of the war.

Diagram of the causes of mortality in the Army in the East

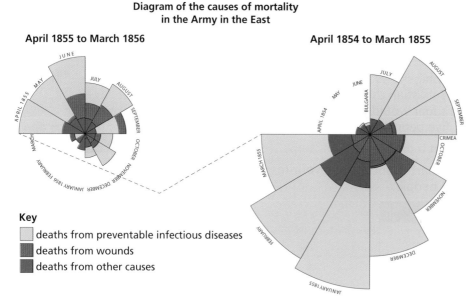

April 1855 to March 1856

April 1854 to March 1855

Key
- deaths from preventable infectious diseases
- deaths from wounds
- deaths from other causes

- Discuss what this graph is showing. Write down some conclusions from this graph.
- Doctors recommended new ways for trying to prevent infection in February 1855. What effect did these changes have on the number of deaths?

4 **Real data question** The infographic shows whether people think the world is getting better or worse in different countries.

Thinking generally about the world, all things considered, do you think the world is getting better or worse, or neither getting better nor worse? 18 235 adults in the following countries. Results ordered by net score (% getting better minus % getting worse)

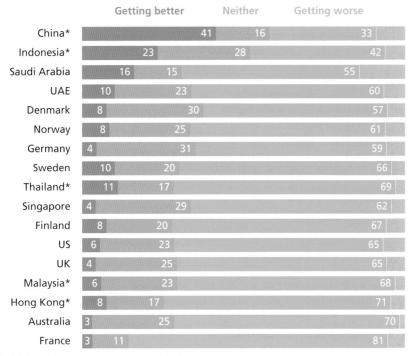

*Weighted to be representative of online population

Source: Yougov

a) Write down the percentage of adults in France who thought the world was getting worse.

b) Compare the percentages of people who thought the world was getting better in Indonesia and Finland.

c) Suggest what the grey regions of the graph might represent.

5 **Real data question** The infographic shows the population sizes of Russia, Nigeria, Pakistan and Indonesia for 2011 and population estimates for 2031 compared to their land areas.

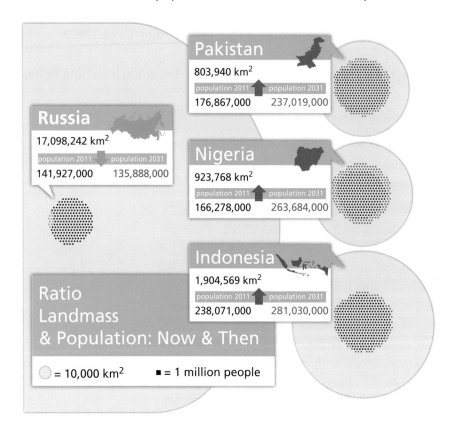

Source: © Design and concept by Kate Snow Design for Bits of Science

a) Write down which one of the four countries has a population that is forecast to decrease.

b) Write down what each small black dot represents.

c) Find the percentage increase in the population of Indonesia between 2011 and 2031.

d) Find the predicted number of people there will be per km² in Russia in 2031.

e) Find which of the countries is the most densely populated. Show how you decided.

Thinking and working mathematically activity

Japan's population by age group, m

■ Male ■ Female

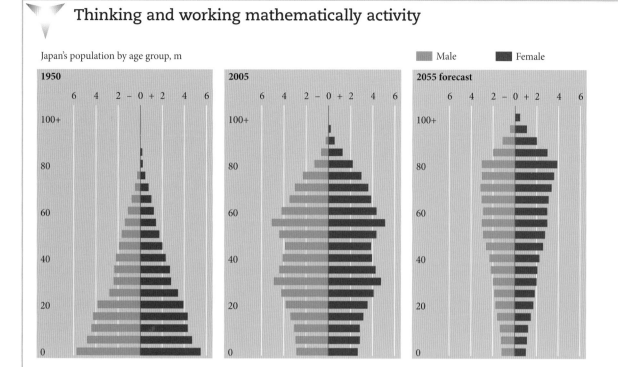

The diagrams show Japan's changing population from 1950 to 2055.

Discuss and comment on how the population of Japan has changed and is forcasted to change between 1950 and 2055.

Investigate similar diagrams showing the population for your a country of your choice. How do the diagrams for your chosen country compare with those for Japan?

20.3 Correlation

Key terms

Two sets of data are **correlated** is there is a relationship between them.
Correlation can be **positive** or **negative**.

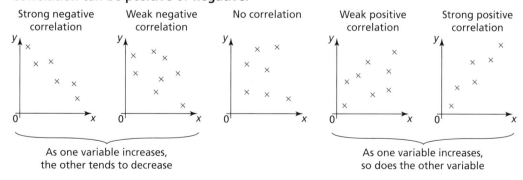

A **conjecture** (or **hypothesis**) is an idea that you want to investigate by collecting data. For example, the cost of a house and the number of bedrooms are positively correlated.

Interpolation uses a line of best fit to estimate a value that is within the data points.
Extrapolation uses a line of best fit to estimate a value outside the plotted data points.

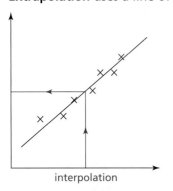

interpolation

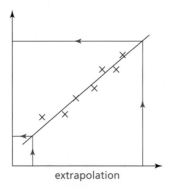

extrapolation

You cannot assume that the data will always follow the same pattern, so **extrapolation** can be unreliable.

Causation is when a change in one variable **causes** a change in the other. For example, increasing the amount of light a plant receives will cause it to grow taller.

Not all variables which are correlated have a relationship that involves causation. For example, a shop may sell sun hats and ice creams. There may be correlation between the number of sun hats sold and the number of ice creams sold. But selling more sun hats does not cause more ice creams to be sold.

 Thinking and working mathematically activity

With regard to the time taken to run a 100 metre race, write down one or two variables that would display:

- positive correlation
- negative correlation
- no correlation

In each case explain why the variable you selected would give the type of correlation.

Worked example 3

The scatter graph shows the arm length and the height of some students.

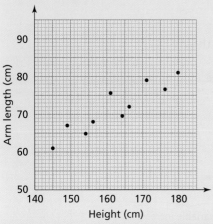

a) Describe the correlation between height and arm length.

b) A different student is 158 cm tall. Use your graph to estimate the student's arm length.

c) Comment on how accurate your estimate is likely to be.

a) There is strong positive correlation between height and arm length. This means that taller people tend to have longer arms than shorter people.

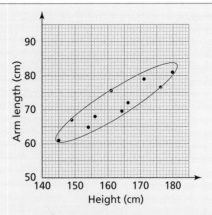

The points can be enclosed by an ellipse which:

- increases – this shows positive correlation (as one variable increases so does the other)
- is narrow – the points are close to a straight line so the correlation is strong.

b)

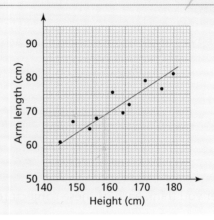

Worked example 4

Rizwana has this conjecture:

Students perform better in tests if they sleep longer at night.

She collects data from students in her class and draws a scatter graph.

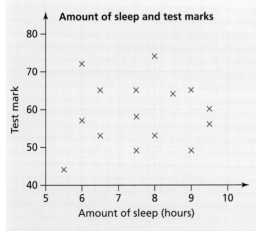

Comment on whether or not Rizwana's data supports her conjecture.

The scatter graph shows no correlation. So Rizwana's data does not support her conjecture.	There is neither an increasing trend nor decreasing trend in the data. If Rizwana's conjecture were true, the scatter graph should be showing positive correlation.	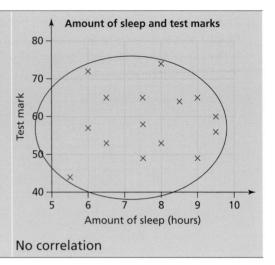

No correlation

Exercise 3

1 Write down the type of correlation shown in each scatter graph.

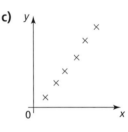

a)   b)   c)   d)

2 Decide whether you think each of these conjectures is likely to be true or false.

a) There is a negative correlation between the number of tickets sold to a tennis match and the amount of money taken.

b) Ice cream sales and temperature are positively correlated.

c) There is no correlation between the wealth of a country and how long people live.

d) The amount of time it takes a child to read a book is negatively correlated with the age of the child.

Discuss

Discuss your answers to question 2 with a partner.

Do you agree on your answers?

---

3 Eight dancers take part in a dance competition. Three judges give each dancer a mark. The scatter graphs show the marks.

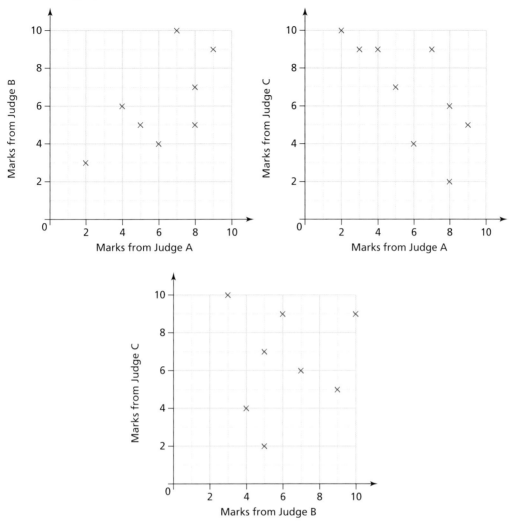

a) Which two judges' marks show weak positive correlation?

b) What can you conclude about the marks awarded by judges A and C?

4 Describe the correlation shown by each graph below.

Decide whether a change in one variable will cause a change in the other. Give reasons for your answers.

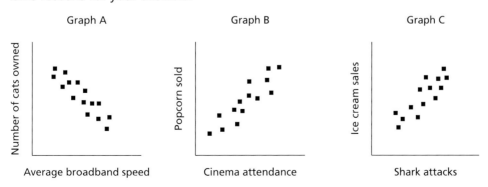

5 Xander sells ice pops. He records the number of ice pops he sells and the temperature on 7 days.

Temperature (°C)	15	17	19	21	25	27	30
Number of ice pops	18	31	33	45	48	58	59

a) Draw a scatter graph to show this information. Describe the strength of the correlation.

b) Draw a line of best fit on your scatter graph.

c) Use your line of best fit to predict the number of ice pops Xander will sell when the temperature is 23 °C.

d) Xander also wants to use the scatter graph to predict the number of ice pops he will sell when the temperature reaches 35 °C. Explain why this may not give a result that is accurate.

6 A teacher records the time it takes 12 students to run a race. She also records the number of skips each student can do in 1 minute. Her results are in the table.

Number of skips	29	33	70	37	38	42	54	46	51	52	63	78
Time (seconds)	210	225	194	212	208	201	195	210	188	207	182	174

a) By first drawing an appropriate graph, describe the correlation between the number of skips and time.

b) A different student managed 60 skips in 1 minute. Use your graph to estimate it will take her to run the race. Show how you worked out your estimate.

7 Students take tests in Maths and Science.

Mishal makes this conjecture:

Students with higher marks in Science also tend to have higher marks in their Maths test.

She collects data from some of her class and draws this scatter graph.

Comment on Mishal's conjecture.

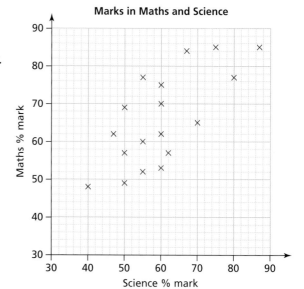

Marks in Maths and Science

8 Laura makes this conjecture:

There is a negative correlation between the number of pieces of fruit a person eats per day and the number of unhealthy snacks eaten.

She collects data from 10 people and displays it in this scatter graph.

Write down whether the data supports Laura's conjecture or not.

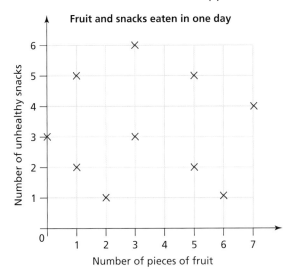

Consolidation exercise

1 The dual bar graph shows the favourite types of film for a sample of boys and girls.

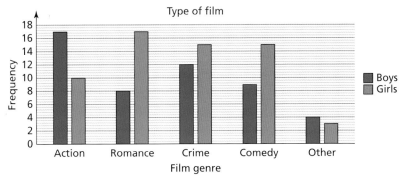

a) Which type of film was the most popular overall?

b) Find the number of girls in the sample.

c) Draw a compound bar chart to display the data.

d) Write down one thing that a dual bar chart shows more clearly than a compound bar chart.

2 The pie chart shows how a sample of 400 people travelled to get to their last holiday.

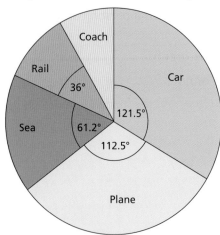

a) Find how many people travelled by plane.

b) Draw a bar chart to show the same data.

c) Write down one advantage that a bar chart has compared with a pie chart.

3 Real life data The diagram shows the ages of people living in Mongolia in 2020.

Mongolia: Population in thousands in 2020

MALES		FEMALES		AGE GROUP
7		13		80 +
25		35		70 – 79
69		89		60 – 69
146		168		50 – 59
208		215		40 – 49
283		281		30 – 39
251		249		20 – 29
253		248		10 – 19
373		364		0 – 9

400 300 200 100 0 100 200 300 400
population (thousands)   population (thousands)

Source: *From World Population Prospects 2019, Online Edition. Rev. 1., by Department of Economic and Social Affairs, Population Division (2019). Copyright © 2019, United Nations. Reprinted with the permission of the United Nations*

a) Write down the number of females aged 30 – 39.

b) Find the total number of people in Mongolia aged under 20.

c) Find the difference between the number of males and females aged 50 – 59.

4 **Real life data** The infographic shows the relationship between carbon dioxide emissions per person and income per person for countries in Asia in 2014.

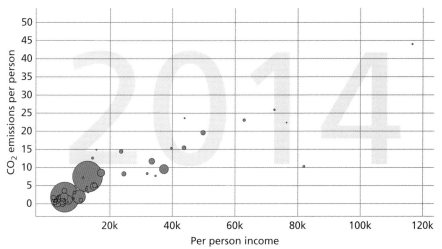

Source: Based on free material from GAPMINDER.ORG, CC-BY LICENSE.

a) Suggest a reason why some of the circles are large and some circles are small.

b) Describe the relationship between carbon dioxide emissions per person and income per person.

5 Andy is a teacher. He has this conjecture:

Students who attend more lessons perform better on tests than students who miss lessons.

He collects data from students in his class and draws a scatter graph.

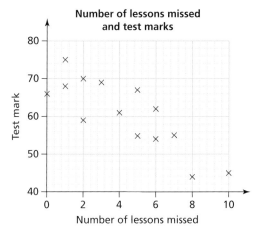

Comment on whether Andy's data support his conjecture or not.

Give a reason for your answer.

6 Here are six scatter graphs.

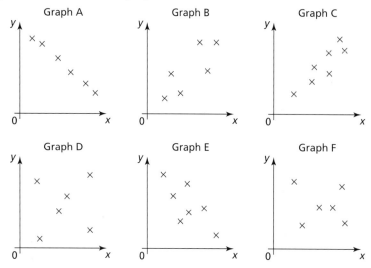

Graph A Graph B Graph C

Graph D Graph E Graph F

Copy and complete this table by writing each scatter graph in the correct column.

Negative correlation	No correlation	Positive correlation

7 A group of students take an exam in Computer Science and Spanish.

The table shows the marks they get in each subject.

Computer Science mark	45	68	74	82	34	52	54	95	85	60
Spanish mark	38	55	63	77	45	63	48	72	84	69

a) Draw a scatter graph to show the information.

b) Describe the correlation between the mark in Computer Science and the mark in Spanish.

c) Draw a line of best fit on the scatter graph. Use it to estimate the mark in Computer Science for a student who got 60 marks in Spanish.

End of chapter reflection

You should know that...	You should be able to...	Such as...
Statistical information can be presented in different forms.	Change data shown in one statistical graph into a different graph.	Draw a pie chart to show the favourite fruits of adults.
An infographic is a visual representation of data presented in an easy to understand form.	Interpret infographics and make calculations based on the data.	Use the infographic to find the percentage of food that was supplied from outside of the UK and the EU. Percentage of food supplied to the UK 2% ⌐1% 4% 4% 4% 4% 50% 30% ■ UK ■ EU ■ Africa ■ North America ■ South America ■ Asia ■ Rest of Europe ■ Australasia

A scatter graph can show positive or negative correlation (or no correlation at all).	Describe the type of correlation shown on a scatter graph.	Describe the type of correlation shown on this scatter graph. 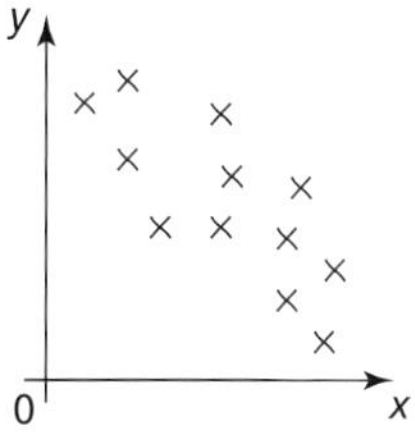
Data is collected to see whether it shows support or casts doubt on an initial conjecture.	Comment on whether or not data shown as a table, graph or diagram gives support to a conjecture.	A teacher makes this conjecture: *The marks scored by students in two tests will be positively correlated.* The test marks are shown in the scatter graph. 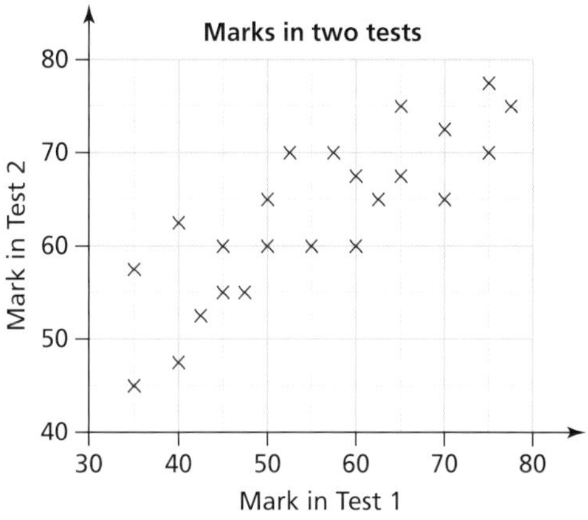 Comment on the teacher's conjecture.
Distributions can be compared by examining the shape of their graphs or by comparing summary measures.	Compare two or more distributions and relate conclusions to the original question.	Nicki wants to compare how far students at two schools live from their school. She collects this data. |	School 1	School 2
Median	1.5 km	3.2 km
Mean	1.7 km	4.1 km
Range	6.5 km	13.2 km | Make some conclusions about how far students live from these two schools.
Correlation does not mean causation.	Decide whether a relationship between two variables is likely to be causal or not.	A scatter graph shows strong positive correlation between sunglasses sales and flip flop sales. Decide whether buying more flip flops will cause the sales of sunglasses to rise.

| Interpolation uses a line of best fit to estimate a value that is within the data points. Extrapolation uses a line of best fit to estimate a value outside the plotted data points. | Extrapolated data may not be reliable as it is outside of the data set. | Freddy wants to use the scatter graph to estimate the money saved by someone who earns a salary of $30 000.

Explain why this may not be a reliable estimate.

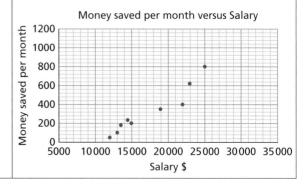 |

Ratio and proportion

You will learn how to:

- Use knowledge of ratios and equivalence for a range of contexts.
- Understand the relationship between two quantities when they are in direct or inverse proportion.

Starting point

Do you remember…

- how to simplify a ratio?

  For example, write the ratio 12 : 48 : 20 in its simplest form.

- how to share a quantity in a ratio?

  For example, share $48 in the ratio 5 : 7

- how to use equivalence to compare ratios and solve problems?

  For example, which is the better value: 400 g of pasta for $1.60, or 0.45 kg of pasta for $1.80?

This will also be helpful when…

- you learn to solve more complex ratio problems
- you learn to solve direct and inverse proportion problems using algebra.

21.0 Getting started

Kosal travels 3 km to school. Make a table showing his speed and travel time (in hours or fractions of an hour) when he:

- walks to school at 3 km/h
- walks to school at 4 km/h
- jogs to school at 6 km/h

- runs to school at 9 km/h
- cycles to school at 12 km/h
- travels to school by bus at 24 km/h.

Look for relationships between quantities in your table. For example:

What happens to the time if the speed:

- doubles
- is multiplied by 3
- is multiplied by 6
- halves?

What happens to the speed if the time:

- doubles
- is multiplied by 3
- is multiplied by 6
- halves?

Kosal's speed and time, for a particular journey, are 'inversely proportional'.

Write rules describing:

- what happens to the time when the speed is multiplied by a number
- what happens to the speed when the time is multiplied by a number
- the product of the speed and the time.

Key terms

A **proportion** shows the size of a part relative to the whole. It is usually expressed as a fraction.

Worked example 1

Lars and Khadija share some books in the ratio 5 : 1. Lars gets 12 more books than Khadija.
Find the total number of books.

12 ÷ 4 = 3 books in each share	Lars gets four more shares than Khadija. These four shares equal 12 books. Find the number of books in one share.	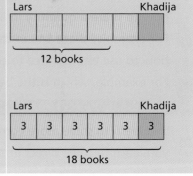
Total shares = 1 + 5 = 6 Total number of books = 6 × 3 <div align="right">= 18</div>	Multiply this by the total number of shares, 6.	

▼ Thinking and working mathematically activity

To make a type of biscuit, the ratio of flour to butter used is 3 : 2. The ratio of butter to sugar is 3 : 4.
Draw similar diagrams to find the ratio $a : b : c$

The ratios 3: 2 and 3: 4 can be represented by these diagrams:

The same ratios can also be represented by these diagrams (think about why):

The combined ratio of flour to butter to sugar can be found by combining the diagrams like this:

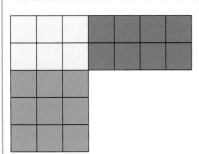

The ratio of flour to butter to sugar is 9 : 6 : 8

Draw similar diagrams to find the ratio $a : b : c$ if:

- $a : b = 3 : 2$ and $b : c = 4$
- $a : b = 4 : 3$ and $b : c = 4 : 5$

Exercise 1

1 A family spends money on food and travel in the ratio 5 : 2.
In one month, they spend $80 on travel.

Find the amount they spend on food in that month.

2 Atanas, Briony and Chathuri share some money in the ratio 2 : 3 : 5. Briony gets $27.

Find the total amount of money shared.

3 Tatiana and Omid share some grapes in the ratio 7 : 3. Tatiana gets 20 more grapes than Omid.

Find how many grapes Omid gets.

4 Two numbers are in the ratio 3 : 2. One of the numbers is 0.6

Find two possible values for the other number.

5 There are 140 visitors in a museum. There are 20 more adults than children.

Find the ratio of adults to children in the museum, in its simplest form.

6 In a game, Mel, Nadia and Olaf score points in the ratio 4 : 3 : 1
The sum of Mel's points and Nadia's points is 24 more than Olaf's points.

Find the number of points Nadia scores.

7 School A has 150 computers and 900 students. School B has 120 computers and 870 students.

Mia calculates that the ratio of computers to students in school A is 1 : 6

She says that the ratio of computers to students in school B is also 1 : 6, because the numbers on both sides are 30 less than the numbers for school A.

School A	School B
Computers : students	Computers : students
150 : 900	**150 : 900**
1 : 6	−30 ⤵⤴ −30
	120 : 870
	Ratio = 1 : 6

Is Mia correct? Explain your answer.

8 The ratios of side lengths of a cuboid are:

length : width = 4 : 3 and width : height = 2 : 5

Gyatso finds the ratio length : height using the three methods shown on the next page.

Method 1	Method 2	Method 3
length : width = $\frac{4}{3}$: 1 width : height = 1 : $\frac{5}{2}$	length : width = 4 : 3 = 8 : 6 width : height = 2 : 5 = 6 : 15	width : length = 3 : 4 width : height = 2 : 5 If width = 60 cm, length = 80 cm If width = 60 cm, height = 150 cm length : height = 80 cm : 150 cm = 8 : 15
length : height = $\frac{4}{3}$: $\frac{5}{2}$	length : height = 8 : 15	length : height = 8 : 15

Explain his methods. What do they have in common?

9 A bag contains red, blue and green marbles. The ratio of red to blue marbles is 6 : 5. The ratio of blue to green marbles is 2 : 3.

Find the ratio of red to blue to green marbles in the bag.

10 In a school sports team, the ratio of first year students to second year students is 2 : 3. The ratio of second year students to third year students is 2 : 5.

Find the proportion of third year students in the team.

11 100 g of butter is needed to make 12 biscuits. 100 g of sugar is needed to make 18 biscuits.

Jo makes biscuits using 300 g more butter than sugar. Find the number of biscuits she makes.

> **Tip**
>
> First find the ratio of the two ingredients in these biscuits.

> **Discuss**
>
> The ratio of Ann's age to Bal's age is 3 : 4. Ann says, 'Our ages will always be in this ratio.' Explain whether or not she is correct.

21.2 Direct and inverse proportion

Key terms

A **variable** is a quantity that can change. Examples of variables are a person's height, the temperature in a room, and the amount of money in a bank account.

Two variables are in **direct proportion** if they are always in the same ratio. You can also say that the two variables are **directly proportional**.

For example, lengths in centimetres and lengths in metres are directly proportional. They are always in the ratio 100 : 1

Two variables are in **inverse proportion** if their product is always the same. You can also say that the two variables are **inversely proportional**.

For example, some people share a box of 36 chocolates equally. The number of chocolates per person is inversely proportional to the number of people. The two quantities always have a product of 36.

Worked example 2

A toy car travels 50 metres in 30 seconds, moving at constant speed.

Find how far it travels in 12 seconds.

distance : time = 50 : 30 = 5 : 3	The distance is directly proportional to the time. (If the time doubles, the distance doubles, for example.)
	Find the ratio of distance : time in its simplest form. For two directly proportional variables, this ratio is always the same.
5 : 3 = 20 : 12 It travels 20 m in 12 seconds	Multiply both sides by 4 to find the equivalent ratio with 12 as the time part.

Worked example 3

It takes 6 days for 3 builders to build a wall.

Find how long it would take for 9 builders to build the same wall.

6 × 3 = 18	The time is inversely proportional to the number of builders. (If the number of builders doubles, the time taken halves, for example.)
	Find the product of the time and the number of builders.
	For two inversely proportional variables, this product is always the same.
18 ÷ 9 = 2 It takes 9 builders 2 days.	Divide 18 by 9 to find the time.

Worked example 4

12 pencils cost $4.44. Find the cost of 7 pencils.

$4.44 × $\frac{7}{12}$ = $2.59	The number of pencils is directly proportional to the cost, so the ratio is always 12 : 4.44. You could look for the equivalent ratio with 7 on the left.
	However, it is easier to use the multiplier $\frac{7}{12}$ (divide by 12 and multiply by 7) to change from 12 pencils to 7 pencils.

Exercise 2

1–5, 7

1 One of the pairs of quantities is different from the others. State which pair, and explain why.

A	the distance a car travels	the amount of fuel it uses
B	the number of carriages on a train	the number of passengers the train can carry
C	the speed of an aeroplane	the time it takes to complete its journey

2 Each table shows values of a pair of variables. For each table, state whether the two variables are directly proportional, inversely proportional, or neither.

a)

p	q
2	6
3	7
5	9
6	10

b)

r	s
2	30
3	20
5	12
6	10

c)

v	w
2	4
4	16
6	36
8	64

d)

x	y
2	14
4	28
6	42
8	56

3 a) An athlete takes 110 seconds to run around a race track. She then runs around the same track at double the speed. Find how long she will take.

b) An athlete runs a distance in 6 minutes. He then runs three times this distance at the same speed. Find how long he will take.

c) An athlete runs 1000 m in a certain time. She then runs for half of this time at the same speed. Find how far she will run.

4 It takes 3 carpenters 24 hours to make 8 chairs. Find how long it takes for:

a) 3 carpenters to make 2 chairs

b) 6 carpenters to make 8 chairs

c) 6 carpenters to make 2 chairs.

> **Tip**
>
> In questions 3 and 4 check whether your answers seem sensible.

5 If 8 kilometres equal 5 miles, convert:

a) 15 miles to kilometres

b) 32 kilometres to miles

c) 2 kilometres to miles

d) 1 mile to kilometres.

e) **Technology question** Use a spreadsheet program to draw a graph for converting between kilometres and miles. Show distance in kilometres on the y-axis.

6 True or false? If 15 British Pounds = 27 Singapore Dollars:

a) 20 British Pounds = 32 Singapore Dollars

b) 300 British Pounds = 5400 Singapore Dollars

c) 450 British Pounds = 810 Singapore Dollars

d) 200 British Pounds = 360 Singapore Dollars

> **Tip**
>
> It is possible to calculate each part using a single line of calculation. See Worked example 4.

7 Mel and Ben received 84.60 US Dollars in exchange for 64 British Pounds.

To calculate the exchange rate for 1 US Dollar, Mel divides 84.60 by 64. Ben divides 64 by 84.60.

a) Who is correct? Explain your answer.

b) How many British Pounds would they get for 150 US Dollars?

c) If they received 210 British Pounds, how many US Dollars did they exchange?

> **Did you know?**
>
> The resting heart rates of different species of animal are approximately inversely proportional to their average lifespans.

> **Think about**
>
> What is wrong with the question 'Is 3 proportional to 6?'

 Thinking and working mathematically activity

Two quantities, x and y, are directly proportional. When $x = 4$, $y = 6$.

Make a table showing at least five pairs of values of x and y, with the x values no more than 20.

Draw x- and y-axes and plot each pair of values as a point. Draw a suitable line to join the points.

Describe the shape of the graph.

For different pairs of directly proportional variables, describe:

- what could be different about their graphs
- what will always be the same about their graphs.

Consolidation exercise

 One of the pairs of quantities is different from the others. State which pair, and explain why.

A	the number of people sharing a task	the time taken to finish the task
B	the number of hours worked	the amount of pay received
C	the time it takes to make one item	the number of items made per day

2 State whether each pair of variables is directly proportional, inversely proportional or neither.

a) the speed of a car and the distance travelled in 1 minute

b) the speed of a runner and the time taken to run 100 metres

c) the number of tickets for a show and the cost per ticket

d) the side length of a square and the perimeter of the square

3 A gardener plants red, yellow, white and pink roses in the ratio 5 : 2 : 3 : 1.
He plants 45 yellow and white roses in total. How many roses does he plant altogether?

4 A scarf in Tunisia costs 75 Tunisian Dinar. The exchange rate is \$2 = 5 Tunisian Dinar.

Jasmine has \$28. Can she buy the scarf? Show your working.

5 True or false? If x is directly proportional to y, then y is inversely proportional to x.

Explain your answer.

6. Some people share a box of chocolates equally. Each person gets 6 chocolates.

Three times as many people share an identical box of chocolates.
Find how many chocolates each person gets.

7 Benoit, Chris and Daumaa share \$70. The amounts Benoit and Chris get are in the ratio 2 : 3.
The amounts Chris and Daumaa get are in the ratio 4 : 5

Find how much Daumaa gets.

End of chapter reflection

You should know that...	You should be able to...	Such as...
You can solve ratio problems using knowledge of one part of the ratio, or the difference between two parts.	Solve ratio problems in which you are given a ratio and the size of one of the parts, or the difference between parts.	Jake and Krishna share some apples in the ratio 5 : 2. Jake gets 21 more apples than Krishna. Find the total number of apples.
If two variables are directly proportional, they are always in the same ratio.	Identify variables that are directly proportional. Solve problems involving directly proportional variables.	True or false? The number of matches won by a football team is directly proportional to the number of goals the team scores. 20 workers can make 40 items in 5 hours. Find how many items can be made by 5 workers in 10 hours.
If two variables are inversely proportional, their product is always the same.	Identify variables that are inversely proportional. Solve problems involving inversely proportional variables.	True or false? The time taken to travel a fixed distance is inversely proportional to the speed. Some people share a box of chocolates equally and each person gets 4 chocolates. Twice as many people share an identical box of chocolates. Find how many chocolates each person gets.

22 Relationships and graphs

You will learn how to:

- Use knowledge of coordinate pairs to construct tables of values and plot the graphs of linear functions, including where y is given implicitly in terms of x ($ax + by = c$), and quadratic functions of the form $y = x^2 \pm a$.
- Understand that straight-line graphs can be represented by equations. Find the equation in the form $y = mx + c$ or where y is given implicitly in terms of x (fractional, positive and negative gradients).
- Understand that a situation can be represented either in words or as a linear function in two variables (of the form $y = mx + c$ or $ax + by = c$), and move between the two representations.
- Read, draw and interpret graphs and use compound measures to compare graphs.

Starting point

Do you remember...

- how to plot a graph of a linear function?

  For example, draw the graph of $y = 3x - 4$.
- how to recognise the gradient and y-intercept of a linear function from its equation?

  For example, state the gradient and y-intercept of $y = 5x + 2$
- how to represent a relationship as a linear function in the form $y = mx + c$?

  For example, the cost of hiring a bike is a fixed cost of $20 and an additional charge of $30 per day. Write down a function to show the cost, $\$C$, of hiring a bike for d days.
- how to draw and interpret graphs in real-life contexts?

  For example, can you draw distance-time graphs for 2 cars and use the graphs to find out what time one car overtakes the other car?

This will also be helpful when...

- you learn how to find acceleration from a speed-time graph
- you draw exponential growth and decay graphs.

22.0 Getting started

The tortoise and the hare

Dennis the hare and Speedy the tortoise are having a race along a 2.7 km track.

Speedy covers the distance in 15 minutes. He moves at a steady speed and does not stop.

Dennis is confident that he can win. He speaks with his friends for 2 minutes before starting to move. It takes him 6 minutes to run the first 1.8 km. Then he stops to sleep for 5 minutes. He completes the rest of the race at the same speed as he was travelling before his rest.

- Plot this information on a time-distance graph.
- Find who won the race.

Chapter 22: Relationships and graphs   **255**

Key terms

You can write all **linear functions** in the form $y = mx + c$, where m and c are constants, or in the form $ax + by = c$, where a, b and c are constants. The form $ax + by = c$ is known as **implicit** form as it does not directly tell you how to find the value of y from a given value of x.

Functions like $y = x^2 + 5$ and $y = x^2 - 3$ are examples of simple **quadratic functions**.

You can draw the graph of a function by constructing a table of values. You can use the coordinate pairs from the table to plot the graph of the function.

The graph of a quadratic function is a curve. This curve is called a **parabola.**

Worked example 1

a) Copy and complete the table of values for $x + 2y = 5$.

x	−2	0	2	4
y				

b) Draw the graph of $x + 2y = 5$.

---

a)

x	−2	0	2	4
y	3.5	2.5	1.5	0.5

Complete the table by substituting the values of x into $x + 2y = 5$.

For example,

when $x = -2$,　　then　　$-2 + 2y = 5$

$$2y = 7$$

$$y = 3.5$$

---

b)

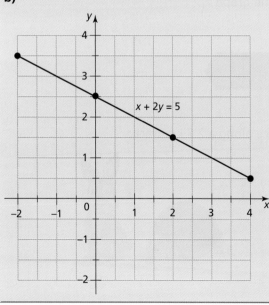

Read the coordinate pairs from your table:

$(-2, 3.5)$, $(0, 2.5)$, $(2, 1.5)$ and $(4, 0.5)$

Plot the coordinate pairs.

Draw a straight line through the points.

Worked example 2

a) Find the points where the graph of $3x - 5y = 15$ intercepts the y-axis and the x-axis.

b) Use these intercepts to draw the graph of $3x - 5y = 15$

a)

When $x = 0$:	$3(0) - 5y = 15$	
	$-5y = 15$	
	So	$y = -3$

When $y = 0$: $\quad 3x - 5(0) = 15$

$\qquad\qquad\qquad 3x = 15$

$\qquad\quad$ So $\qquad x = 5$

The intercepts are at $(0, -3)$ and at $(5, 0)$

To find where the graph intercepts the y-axis, substitute $x = 0$. Then solve to find the value of y.

To find where the graph intercepts the x-axis, substitute $y = 0$. Then solve to find the value of x.

b)

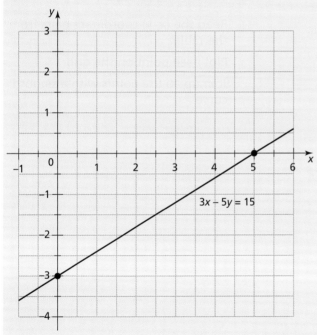

Plot the points $(0, -3)$ and $(5, 0)$.

Draw the line that passes through the two points.

Worked example 3

a) Copy and complete the table of values for $y = x^2 - 5$

x	− 3	− 2	− 1	0	1	2	3
y							

b) Draw the graph of $y = x^2 - 5$

a) $y = x^2 - 5$

x	− 3	− 2	− 1	0	1	2	3
y	4	− 1	− 4	− 5	− 4	− 1	4

Complete the table by substituting each value of x into the equation

$y = x^2 - 5$

For example,

when $x = -3$, $\quad y = (-3)^2 - 5$
$$= 9 - 5$$
$$= 4$$

when $x = -2$, $\quad y = (-2)^2 - 5$
$$= 4 - 5$$
$$= -1$$

b) Plot the points and draw the graph.

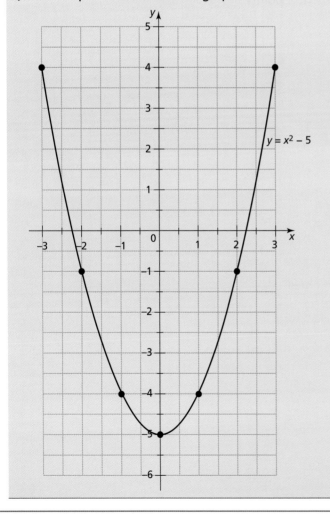

Read the (x, y) coordinate pairs from the table.

Plot each point on the grid.

Join up all the points with a smooth curve.

Exercise 1

1 **a)** Make y the subject of the linear function $2x + y = 6$.

   b) Copy and complete the table of values for this function.

x	−2	0	2	4
y				

   c) Draw the graph of the function. Use an x-axis from −2 to 4 and a y-axis from −2 to 10.

2 Here are equations of two linear functions: $x + y = 4$ and $2x + y = 7$.

   a) Copy and complete a table of values for each function.

x	−1	0	1	2
y				

   b) Plot the graphs of both functions on the same grid. Use an x-axis from −1 to 2 and a y-axis from −1 to 9.

3 Draw the graph of each function. In each case, begin by completing a table of values using x values equal to −2, 0, 2 and 4.

   a) $3x + y = 6$ **b)** $y - 2x = 3$ **c)** $x + 2y = 4$

4 **a)** Copy and complete the table of values for the function $3x + 2y = 9$.

x	−1	0	1	2
y				

   b) Draw the graph of the function $3x + 2y = 9$.

5 **a)** Find the points where the graph of the function $3x + 4y = 12$ intercepts the x- and y-axes.

   b) Use the values of these intercepts to draw the graph of $3x + 4y = 12$

6 Use the intercept method to draw the graph of each of the following linear functions:

   a) $2x + 5y = 10$ **b)** $5y - 2x = 10$ **c)** $3y - 2x = 9$

7 Ida says that the graph drawn on this grid has the equation $x + 2y = 5$.

Is she correct? Explain your answer.

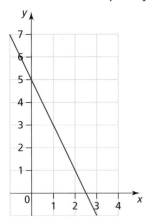

▼ Thinking and working mathematically activity

Technology question Use graph drawing software to draw graphs of functions of the form $ax + by = c$. You should try drawing some graphs with positive coefficients and some graphs with negative coefficients.

Find a rule that will predict when a graph of a function of the form $ax + by = c$

• will have a y-intercept that is positive

• will have a gradient that is positive.

Use your rules to help you place the graphs of the following functions in the correct positions in the table.

$3x - y = 9$	$6x + 5y = 7$	$-2x + 5y = 11$
$x - 3y = -5$	$-2x - 5y = 9$	$-x + 6y = -4$

	Positive gradient	Negative gradient
Positive y-intercept		
Negative y-intercept		

---

8 a) Copy and complete this table of values for the function $y = x^2$.

x	-3	-2	-1	0	1	2	3
y							

b) Draw the graph of $y = x^2$ for values of x between -3 and 3.

9 a) Copy and complete this table of values for the function $y = x^2 + 1$.

x	-3	-2	-1	0	1	2	3
y							

b) Draw the graph of $y = x^2 + 1$ for values of x between -3 and 3.

10 a) Copy and complete this table of values for the function $y = x^2 - 3$.

x	-3	-2	-1	0	1	2	3
y							

b) Draw the graph of $y = x^2 - 3$ for values of x between −3 and 3.

> **Discuss**
>
> Look at the graphs in questions 9 and 10. What is the same and what is different between them?

22.2 Equation of a straight-line graph

Worked example 4

Find the equation of each line.

a)

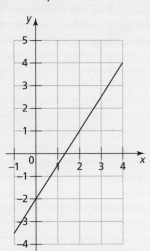

b)

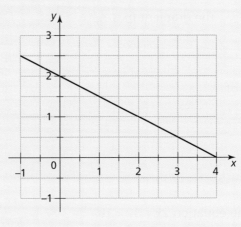

> **Tip**
>
> Remember that the equation of a straight line can be written in the form $y = mx + c$, where
> - m is the gradient
> - c is the y-intercept.

a) The equation of a straight line is $y = mx + c$ $m = 1.5$ $c = -2$ So, the equation of the line is $y = 1.5x - 2$	The y-intercept of this graph is -2. So $c = -2$ To find the gradient, identify two points that the line passes through. Find the change in the y-coordinates and the change in the x-coordinates. The gradient can be found using gradient $= \dfrac{\text{increase in } y\text{-coordinates}}{\text{increase in } x\text{-coordinates}}$ The gradient is $m = \dfrac{6}{4} = 1.5$	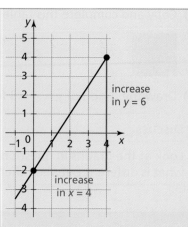
b) The equation of a straight line is $y = mx + c$ $m = -\dfrac{1}{2}$ $c = 2$ So, the equation of the line is $y = -\dfrac{1}{2}x + 2$	The y-intercept is 2. So $c = 2$ This graph goes **down** by 1 when the x-values increase by 2. The gradient m is $\dfrac{-1}{2} = -\dfrac{1}{2}$	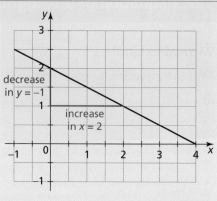

Worked example 5

A straight line has the equation $2x + 3y = 18$.

Find the gradient and y-intercept.

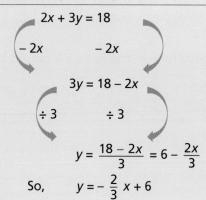$2x + 3y = 18$ $-2x \qquad\qquad -2x$ $3y = 18 - 2x$ $\div 3 \qquad\qquad \div 3$ $y = \dfrac{18 - 2x}{3} = 6 - \dfrac{2x}{3}$ So, $\quad y = -\dfrac{2}{3}x + 6$	Rearrange the equation so that it is in the form $y = mx + c$. Use the balance method – what you do to one side, you must also do to the other. Write as two separate fractions. Change the order so that it is in the form $y = mx + c$.

$m = -\dfrac{2}{3}$ so, the gradient is $-\dfrac{2}{3}$.

$c = 6$, so the coordinates of the y-intercept are $(0, 6)$.

m is the value of the gradient and c is the value of the y-intercept.

1 Write these equations in the form $y = mx + c$ and then find the gradient and y–intercept.

a) $x = y + 5$

b) $y - x = -3$

c) $\frac{x}{2} + y = 7$

d) $2y = 4x + 6$

e) $2y + 6x = 2$

f) $3y - 6x = 12$

g) $\frac{1}{2}y = x + 1$

h) $2y + x = 3$

i) $3y - 3x = 1$

j) $x - 4y = 8$

k) $3x - 2y = 6$

l) $2x - \frac{1}{2}y = 3$

2 In each part, find which equation is the odd one out. Give a reason for each choice.

a) $y = 2 - 5x$ $x - 5y = 3$ $5x + y = 6$ $3y + 15x = 7$ $2y = 11 - 10x$

b) $y = 5x + 2$ $x - 3y = -15$ $5x + y = 5$ $2y - 4x = 10$ $\frac{1}{2}y + 10x = 2.5$

> **Think about**
>
> Without drawing the graphs, how are the graphs of
> $y = 0.25x - 2$ and $8y - 2x = 7$ the same? How are they different?

3 a) Find the gradient of the line.

b) Where does the line intersect the y-axis?

c) Write the equation of the line in the form $y = mx + c$.

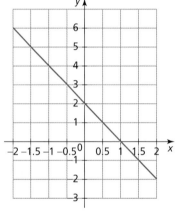

4 Find the equation of each line. Give each equation in the form $y = mx + c$.

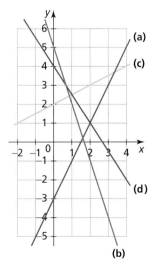

A straight line passes through the points (0, –2) and (–4, –3).

Which of these are correct equations for the line? Show how you made your choices.

- $y = -\dfrac{1}{4}x - 2$
- $4y - x = -8$
- $-4y + x = -2$
- $-4x - 3y = -2$
- $y = \dfrac{1}{4}x - 2$

Decide whether each of the following points lie on the line or not. Show how you decided.

- (16, 2)
- (–5, –4)
- (8, 0)
- (–20, –7)
- (32, 4)

5 Draw a pair of coordinate axes. Number the x-axis from –1 to 3 and the y-axis from –5 to 10.

A line has equation $y = 3x + 1$.

a) What is the y-intercept? Plot the y-intercept on your axes.

b) Explain why the gradient is 3.

c) Start at the y-intercept from part a. Draw a straight line with gradient 3.
Go across 1 unit and up by 3 units. Extend your line to both edges of the grid.

6 Use the method in question 5 to plot these graphs.

a) $y = 2x + 2$ **b)** $y = x - 2$ **c)** $y = -2x + 8$ **d)** $y = \dfrac{1}{3}x + 4$

7 Valma draws the graph of $y = -\dfrac{1}{2}x + 2$.

She starts at (0, 2). Then she moves 1 unit across and 2 units down.

What mistake has Valma made? Show how to
draw the graph correctly.

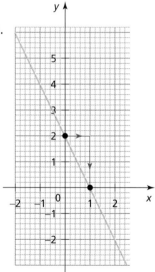

8 **a)** Match each equation to the correct graph.

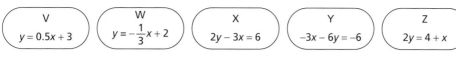

V	W	X	Y	Z
$y = 0.5x + 3$	$y = -\dfrac{1}{3}x + 2$	$2y - 3x = 6$	$-3x - 6y = -6$	$2y = 4 + x$

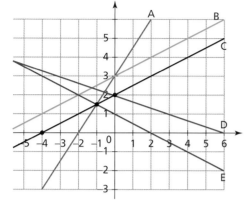

b) **Technology question** Check your answers by plotting the fives lines using graph drawing software.

22.3 Forming linear functions

Worked example 6

Whiteboard pens are sold in large packs and small packs. A large pack has 10 pens and a small pack has 4 pens. A shop buys a total of 200 pens.

a) Write a function to represent the situation. Use l to represent the number of large packs the shop buys and s to represent the number of small packs the shop buys.

b) The function $1.20l + 0.80s = 90$ represents the cost (in dollars) of the pens the shop buys. What do the numbers 1.20, 0.80 and 90 represent in this situation?

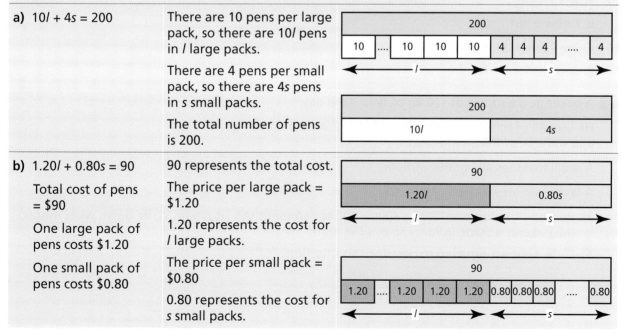

a) $10l + 4s = 200$

There are 10 pens per large pack, so there are $10l$ pens in l large packs.

There are 4 pens per small pack, so there are $4s$ pens in s small packs.

The total number of pens is 200.

b) $1.20l + 0.80s = 90$

Total cost of pens = \$90

One large pack of pens costs \$1.20

One small pack of pens costs \$0.80

90 represents the total cost.

The price per large pack = \$1.20

1.20 represents the cost for l large packs.

The price per small pack = \$0.80

0.80 represents the cost for s small packs.

Tip

In Worked example 6, the letters to be used for the variables (*l* and *s*) were given to you. In some situations, you will need to choose your own letters. You can choose any sensible letters, but you should write down what your variables represent.

Exercise 3

1 At the start of a journey, a plane has 50 000 litres of fuel in its tank. The plane uses 10 litres of fuel per kilometre travelled. Write a formula for the volume, V litres, of fuel in its fuel tank after travelling k kilometres.

2 Ice creams cost $1.10 each. Ice pops cost $0.90 each. Saavi spends $60 buying c ice creams and p ice pops for her party.

Write a function to represent this situation.

3 A coach seats 15 students and a bus seats 45 students. 360 students travel by coach or bus every morning to go to school. Every coach and bus is full.

a) The situation is modelled by the function $15x + 45y = 360$. Which of these statements are true?

   A x = the number of students that travel in one coach

   B y = the number of buses used in the morning

   C x = the number of coaches used in the morning

   D y = the number of students that travel in one bus.

b) Draw a graph of the function $15x + 45y = 360$.

4 John has a piece of wire that is 161 cm long. He uses the wire to make a rectangle with width w cm and length l cm. Which of these functions represents this situation?

   A $w + l = 161$

   B $l \times 2w = 161$

   C $2w + 2l = 161$

   D $w = 161 - l$

5 A baker gets a supply of 120 kg of flour each day.

He uses all of this flour every day by baking some small loaves of bread and some large loaves of bread.

A small loaf uses 400 grams of flour.

A large loaf uses 800 grams of flour.

a) He forms the function $400x + 800y = 120$ to represent this situation. Write down what mistakes the baker has made in forming this function.

b) Form a correct function to describe this situation.

6 Tina wants to spend $400 buying some fabric to make shirts.

Plain fabric costs $4 per metre.

Patterned fabric costs $5 per metre.

Write a function to represent this situation.

You will need to introduce some variables. Remember to write down what they represent.

Thinking and working mathematically activity

Angus makes gold jewellery.

He has 800 grams of gold available. He wants to use all of this gold making bracelets and necklaces.

A gold bracelet requires 40 grams of gold.

A gold necklace requires 72 grams of gold.

* Write down a function that describes this situation.

* Find possible values for the number of bracelets and the number of necklaces he can make. Is there just a single answer?

22.4 Real-life graphs

Key terms

An **exchange rate** is used to convert between two currencies. For example, the exchange rate between US dollars and South African Rand is $1 = 17 Rand. An exchange rate can be used to create a **conversion graph**, for example:

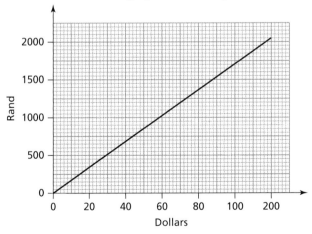

The gradient of a real-life graph represents a **rate of change**.

The gradient of a distance-time graph gives the **speed** of an object.

Worked example 7

Malai is doing a science experiment. She hangs different amounts of mass from a spring and measures the length of the spring. The graph shows the length of the spring, L centimetres, when mass M grams is hanging from it.

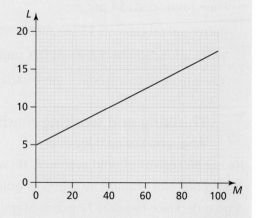

a) Use the graph to find the length of the spring when no mass is hanging from it.

b) Find the gradient of the line.

c) Explain what the gradient represents in this situation.

d) Write a formula for the length, L centimetres, of the spring when mass M grams is hanging from it.

a) $L = 5$ cm	Find the y-intercept, which gives the value of L when $M = 0$.
b)	Use a triangle to find the gradient of the line.

gradient $= \dfrac{\text{change in } y}{\text{change in } x} = \dfrac{10}{80} = \dfrac{1}{8}$

c) The gradient represents the spring's change in length per gram of mass. (Its length increases by $\frac{1}{8}$ cm for each gram added.)	The gradient is $\dfrac{\text{change in length}}{\text{change in mass}}$, with unit cm/g.
d) $L = \frac{1}{8}M + 5$	A straight line has an equation of the form $y = mx + c$, where m represents the gradient and c represents the y-intercept. This line has y-intercept 5 and gradient $\frac{1}{8}$.

Worked example 8

Filipe makes a journey. His journey is shown on the distance–time graph.

a) Find the speed for Stage 3 of Filipe's journey.

b) Find the speed for Stage 2.

c) During which stage of his journey does Filipe travel fastest? Give a reason for your answer.

d) Find the average speed for the entire journey.

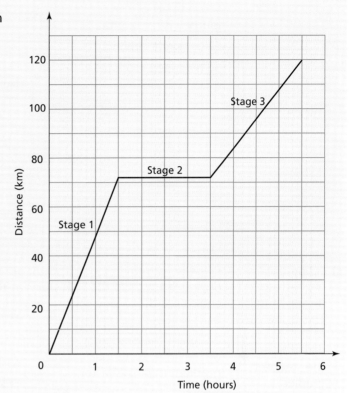

a) Gradient = $\frac{48}{2}$

= 24

The speed for Stage 3 is 24 km/h

The gradient in a distance-time graph represents the speed.

For Stage 3, the gradient is $\frac{48}{2}$ = 24

Include the units of the speed. Here the distance is in km and the time is in hours, so the units of speed here are km/h.

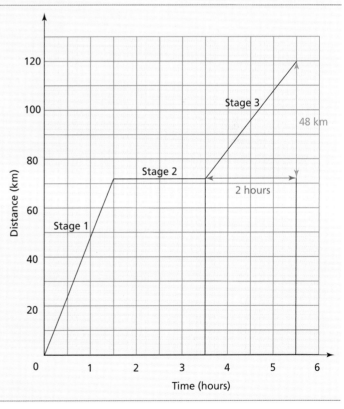

b) Speed = 0 km/h | The graph for the second part of the journey is horizontal.

Filipe is not moving.

So his speed is 0 km/h

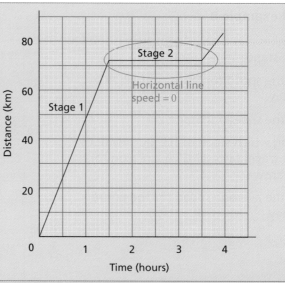

c) Felipe travels the fastest during Stage 1. This is the part of the graph which has the steepest gradient. | Filipe is not moving during Stage 2, so the choice is either Stage 1 or Stage 3.

Compare the gradients of the lines. A steeper gradient means a faster speed. | This diagram shows the line for Stage 3 drawn so that it starts at the origin.

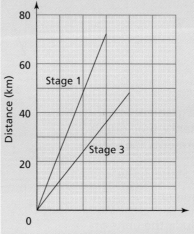

The line for Stage 1 is clearly steeper.

9 Water is draining out of two tanks.

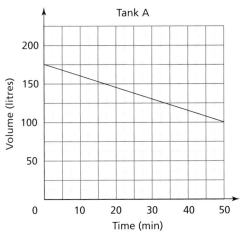

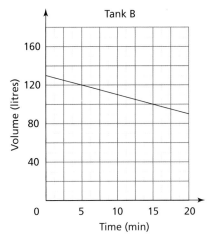

a) Find the rate at which water is draining from Tank A.

b) Compare the rate at which water is draining from the two tanks. Show your working.

c) Jane says that it will take less than 1 hour for Tank B to empty. Is she correct?
Give reasons for your answer.

10 Annie drives to visit a friend. On the way there, she stops to buy petrol.

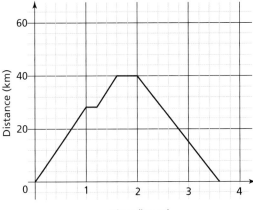

a) How long does Annie stay at her friend's house?

b) Find the speed Annie travelled during the first hour.

c) Find Annie's speed for her journey home from her friend's house.

> **Tip**
>
> In part c of question 10, the speed can be found by finding the gradient of the line. Ignore the negative sign, because Annie's speed will be a positive value.

Thinking and working mathematically activity

This conversion graph can be used to convert between US dollars ($) and Egyptian pounds (EGP).

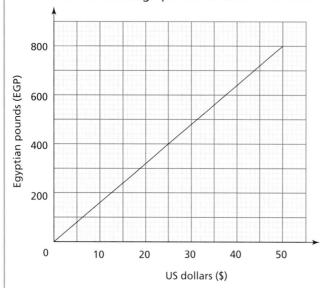

The exchange rate between US dollars and Indian rupees is 1 US dollar = 75 Indian rupees.

- The table shows the cost of some items in the United States, India and Egypt. For each item, decide where it is cheapest.

	Egypt (Egyptian pound RGP)	India (rupees INR)	United States (US $)
Car	1 850 000	7 520 000	96 000
Television	3100	14 500	240

Show clearly how you found your answers.

- Create a conversion graph for converting between Egyptian pounds (RGP) and Indian rupees (INR).

Consolidation exercise

1 Copy and complete this table.

Equation of line	Gradient of line	y-intercept
$y = \frac{1}{2}x + \ldots$	....	8
$y = 5 - x$	....	....
$2y = 3x + \ldots$	....	5
$9x + \ldots y = 6$	−3	....
$2x + 4y = 10$	....	....
$\ldots x + \ldots y = \ldots$	6	−5

2 Here are the equations of four lines:

$\frac{1}{2}y - 4x = 1$ $3y + 4x = 9$ $2y - 4x = 8$ $y - 7x = 5$

Adam says that the line $\frac{1}{2}y - 4x = 1$ has the steepest gradient. Is he correct? Explain why.

3 A party organiser wants to buy exactly 200 balloons for a party. She wants to use a mixture of blue and green balloons. Bags of blue balloons contain 4 balloons. Bags of green balloons contain 5 balloons.

   a) Write a function to represent this situation. Use x = number of bags of blue balloons and y = number of bags of green balloons.

   b) Draw the graph of your function.

4 The graph shows the relationship between temperatures in two different units: degrees Fahrenheit (°F) and degrees Celsius (°C).

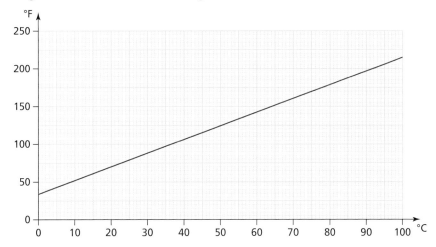

   a) Find the gradient of the graph.

   b) The y-intercept of the graph is 32. Write a formula for converting a temperature in degrees Celsius into degrees Fahrenheit.

> **Did you know?**
>
> In the USA, Belize, the Bahamas, Palau, and the Cayman Islands, temperature is usually measured in degrees Fahrenheit. In the rest of the world, temperature is usually measured in degrees Celsius.

5 a) There are some mistakes in this table of values. Spot the mistakes and correct them.

x	−4	−3	−2	−1	0	1	2	3	4
$y = x^2 -2$	−18	−11	−6	−1	−2	−1	4	7	14

   b) Use the corrected values of the table to draw the graph of $y = x^2 - 2$

6 Simona and Jackson each go on a cycle journey. Information about the first 60 seconds of their journeys is shown on the graph.

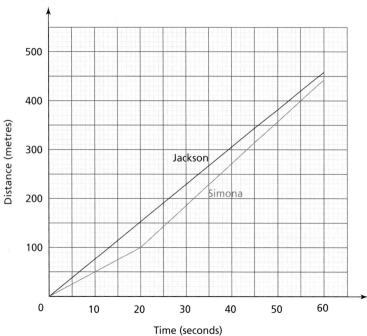

a) Calculate Jackson's speed over the first 60 seconds.

b) Describe the first 60 seconds of Simona's journey.

c) Who was cycling faster 40 seconds after the start? Give a reason for your answer.

End of chapter reflection

You should know that...	You should be able to...	Such as...
A situation can be represented in words or as a linear function in two variables.	Write a function with two variables based on a real-life scenario.	50 people go on a journey. They travel in mini buses and cars. A mini bus seats 10 people. A car seats 5 people. Write a function to represent the situation.
A linear function can be defined implicitly. The graph of a quadratic function has the shape of a parabola.	Draw the graph of linear functions written in the form $ax + by = c$. Draw the graph of quadratics like $y = x^2 + 4$ or $y = x^2 - 9$.	Draw the graphs of a) $x + 3y = 6$ b) $y = x^2 - 7$

Straight-line graphs can be expressed in the form $y = mx + c$ where m is the gradient and c is the y-intercept.	Find the equation of a straight-line graph. Rearrange linear functions to the form $y = mx + c$ to find the gradient and y-intercept.	Find the equation of the line shown. 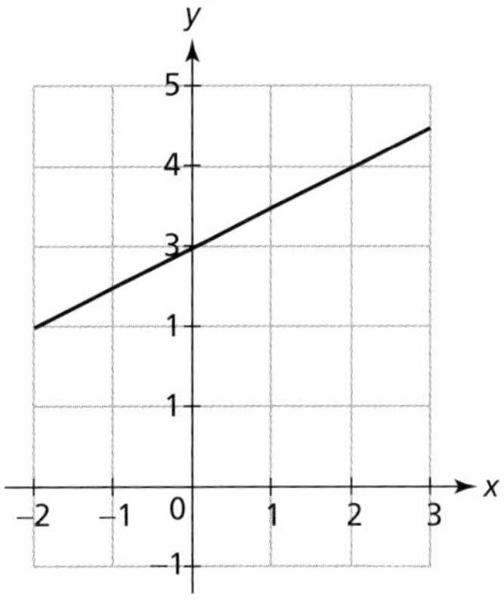 • Write down the gradient and the y-intercept of the line $3y + 4x = 1$
A graph can be used to show a real-life relationship between two variables. Graphs can be used to compare compound measures.	Calculate a compound measure from a real-life graph.	The distance–time graph shows Tom's journey. 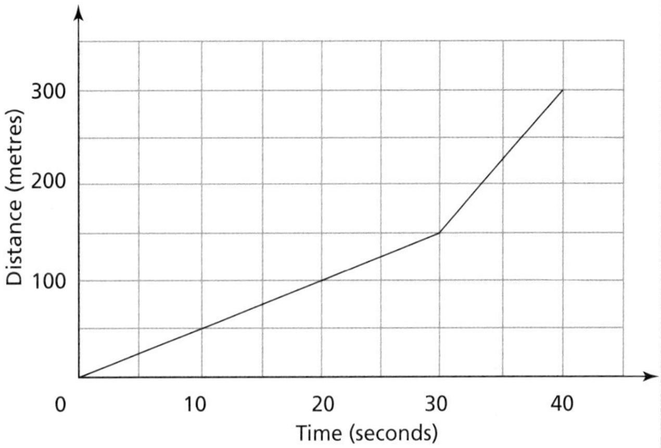 Find Tom's speed for the first 30 seconds of his journey.

23

Probability 2

You will learn how to:

- Identify when successive and combined events are independent and when they are not.
- Understand how to find the theoretical probabilities of combined events.

Starting point

Do you remember…

- how to multiply fractions?

  For example, find $\frac{3}{4} \times \frac{5}{6}$

- that the probability of an event not happening is 1 minus the probability that it does happen?

  For example, find the probability that a biased coin lands on tails if P(heads) = 0.72

- how to list the outcomes for combined events?

  For example, students at a school choose two after-school clubs.
  They choose one of photography, science or tennis and then one
  of football, history or chess. List all the possible club choices.

- how to draw a tree diagram to show all possible combinations of events?

  For example, show all the possible club choices above as a tree diagram.

- how to draw and use a sample space diagram to find probabilities?

  For example, the spinner is spun twice and the scores are added.
  Draw a sample space diagram to show the possible total scores.
  Use it to find the probability that the total score is five.

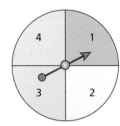

This will also be helpful when…

- you learn how to draw tree diagrams for events that are not independent.

23.0 Getting started

Pedro and Rita play this game.

Pedro takes a counter from a bag and Rita spins a spinner.

They add together the numbers they get.

Pedro's bag

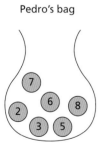

Rita's spinner

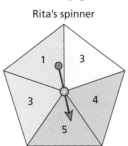

Pedro suggests these rules for deciding the winner:

Pedro wins if the total score is 8, 9, 10 or 11.

Rita wins if the total score is any other number.

- Show that Pedro's rules would not make a fair game.
- Suggest how the rules for deciding the winner could be changed to make a fair game.

23.1 Independent events

Key terms

Two events are called **independent** if they do not affect each other. The probability of one of the events does not depend on whether or not the other happens.

Events that are not independent are called **dependent**.

Worked example 1

a) Toby has a dice and a five-sided spinner.

He throws the dice and spins the spinner.

A is the probability of getting a 5 on the dice.

B is the probability of getting a 3 on the spinner.

Decide whether A and B are independent.

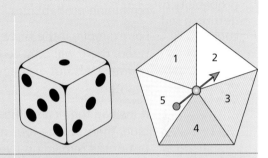

b) Asya has a bag containing four red counters and two blue counters.

She picks a counter from the bag at random.
She does **not** put the counter back in the bag.

She then picks a second counter from the bag.

C is the probability of getting a red counter on the first pick.

D is the probability of getting a red counter on the second pick.

Give a reason why C and D are not independent.

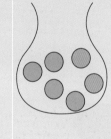

a) A and B are independent.	Events are independent if the probability of the first event happening does not affect the probability of the second event happening. Here, the number that is thrown on the dice will not affect the outcome from the spinner.	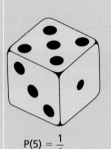 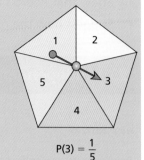 $P(5) = \dfrac{1}{6}$ $P(3) = \dfrac{1}{5}$
b) C and D are not independent because the probability of getting a red counter on the second pick changes depending on the outcome of the first pick.	After the first pick, the counter is not replaced in the bag. This means that on the second pick, one of the counters will be missing. This will affect the remaining probabilities. For example, if the first counter picked is red, there will only be three red counters left out of the remaining five counters. If instead the first counter picked is blue, there will be four red counters left out of the remaining five counters.	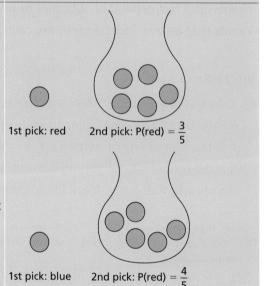 1st pick: red 2nd pick: $P(\text{red}) = \dfrac{3}{5}$ 1st pick: blue 2nd pick: $P(\text{red}) = \dfrac{4}{5}$

Exercise 1

1 Guto travels to work by bus.

A is the probability of missing his usual bus.

B is the probability of being late to work.

C is the probability Guto wears black socks.

a) Write down whether or not A and B are likely to be independent. Give a reason for your answer.

b) Write down whether or not B and C are likely to be independent. Give a reason for your answer.

2 Masha sometimes does some gardening.

P is the probability that it snows on Sunday.

Q is the probability that Masha does some gardening on Sunday.

Decide whether P and Q are likely to be independent or dependent events. Give a reason for your answer.

3 Megan and Nathan take part in the same race.

M is the event that Megan wins the race.

N is the event that Nathan wins the race.

Are the events M and N independent? Give a reason for your answer.

4 Lillia throws an ordinary dice twice. Here are some pairs of events.

A: getting an even number on first throw

B: getting an even number on second throw

C: getting a 6 on the first throw

D: getting an odd number on the second throw

E: getting a 1 on the first throw

F: getting an even number on first throw

G: getting an odd number on the first throw

H: getting a 5 on the first throw

Copy the table below and write each pair of events in the correct column.

Independent events	Dependent events

5 **a)** Jack has five cards numbered 1, 2, 3, 4 and 5 in a bag. He takes a card from the bag, looks at it and then puts in back in the bag. He then takes a second card.

A is getting a 5 on the first pick.

B is getting a 5 on the second pick.

Decide if A and B are independent events. Give a reason for your answer.

b) Jana puts seven coloured balls in a bag: three are green and four are yellow.

She takes a ball from the bag and does **not** replace it. She then picks a second ball.

G is getting a green ball on the first pick.

H is getting a green ball on the second pick.

Decide if G and H are independent events. Give a reason for your answer.

6 George spins a coin nine times and gets nine heads. He then spins the coin a 10th time.

He says, 'I am really unlikely to get a head on the 10th spin.'

Is he correct? Give a reason for your answer.

Thinking and working mathematically activity

An 8-sided spinner is numbered 1, 2, ..., 8.

The spinner is spun twice.

- Write down three pairs of independent events.
- Write down three pairs of events that are not independent.

Share your pairs of events with others.

Think about

....

Amol throws an ordinary red dice and a blue dice. He adds together the two scores.

A: The total score is 12.

B: The red dice shows an odd number.

Are these two events independent?

23.2 Tree diagrams and calculating probabilities

Key terms

....

When two events, A and B, are independent, the probability that both events occur can be found using the **multiplication rule**:

P(A and B) = P(A) × P(B)

Combined events are sometimes shown on a **tree diagram**. When the events are independent:

- the probabilities for each event separately are written on the branches
- the probabilities for combined events can be found by multiplying along the branches.

For example, this spinner is spun twice. The tree diagram looks like this:

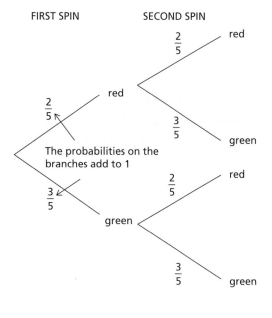

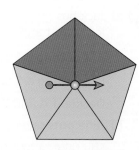

FIRST SPIN   SECOND SPIN

$\frac{2}{5}$ — red   P(red and red) $= \frac{2}{5} \times \frac{2}{5} = \frac{4}{25}$

$\frac{2}{5}$ red

$\frac{3}{5}$ — green   P(red and green) $= \frac{2}{5} \times \frac{3}{5} = \frac{6}{25}$

The probabilities on the branches add to 1

$\frac{2}{5}$ — red   P(green and red) $= \frac{3}{5} \times \frac{2}{5} = \frac{6}{25}$

$\frac{3}{5}$ green

$\frac{3}{5}$ — green   P(green and green) $= \frac{3}{5} \times \frac{3}{5} = \frac{9}{25}$

Total = 1

---

284   Stage 9: Student's Book

Worked example 2

Sara spins a coin and throws a dice.

Find the probability that she gets a tail on the coin and a multiple of 3 on the dice.

P(tail) = $\frac{1}{2}$

P(multiple of 3) = $\frac{2}{6}$ = $\frac{1}{3}$

Spinning a coin and throwing a dice are independent events.

Begin by writing down each probability separately.

The outcomes can be seen in a sample space diagram:

	1	2	3	4	5	6
H	H1	H2	H3	H4	H5	H6
T	T1	T2	T3	T4	T5	T6

P(tail and multiple of 3)

= $\frac{1}{2} \times \frac{1}{3} = \frac{1}{6}$

As the events are independent, the probability of **both** events happening can be found by multiplying.

When the events are combined there are 12 equally likely outcomes.

Sara gets a tail and a multiple of 3 in 2 of these 12 outcomes.

So P(tail and multiple of 3) = $\frac{2}{12} = \frac{1}{6}$

Worked example 3

Katrina spins these two spinners.

a) Draw a tree diagram to show the outcomes and probabilities.

b) Find the probability that she gets a 2 on both spinners.

c) She adds together the numbers she spins to get a total. Find the probability that the total is exactly 4.

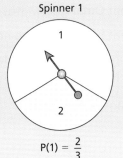

Spinner 1   Spinner 2

P(1) = $\frac{2}{3}$ P(3) = $\frac{1}{4}$

a)

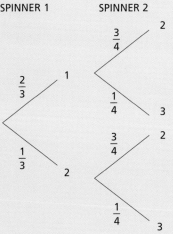

Draw the basic structure for the tree diagram. Write the outcomes at the end of each branch.

Include the probabilities on each branch.

Remember that the probabilities on each branch should add to 1. So

For Spinner 1 : P(2) = 1 − $\frac{2}{3}$ = $\frac{1}{3}$

For Spinner 2 : P(2) = 1 − $\frac{1}{4}$ = $\frac{3}{4}$

b) P(2 on both spinners) =
P(2 on Spinner 1)
× P(2 on Spinner 2)

$= \frac{1}{3} \times \frac{3}{4}$

$= \frac{3}{12}$ or $\frac{1}{4}$

Identify the route through the tree diagram that gives a 2 on both spinners.

The outcomes are independent, so find the probability by multiplying the probabilities along the branches.

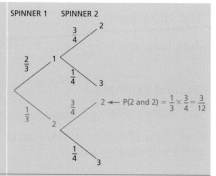

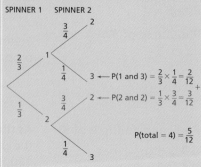

---

c) P(total = 4) = P(1, 3) + P(2, 2)

$P(1, 3) = \frac{2}{3} \times \frac{1}{4} = \frac{2}{12}$

$P(2, 2) = \frac{3}{12}$ (from part **b**)

So P(total = 4) $= \frac{2}{12} + \frac{3}{12} = \frac{5}{12}$

Identify the routes through the tree diagram where the total of the numbers is exactly 4. There are two routes.

Multiply the probabilities along the branches to find the probability of each combined event.

Add these probabilities together to get the overall probability.

Exercise 2

1 The probability that Omar gets Maths homework on Monday is 0.3

The probability that he gets Science homework on Monday is 0.8

Assuming that he gets homework independently, find the probability that he gets both Maths and Science homework on Monday.

2 Torquil has two bags of counters.

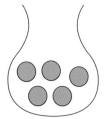 

He takes a counter from each bag at random.

Find the probability that he takes:

a) a blue counter and the number 3.

b) a blue counter and a counter with an even number.

c) a blue counter and a counter with a number greater than 5.

Discuss

Does it matter which spinner is used for the first set of branches of the tree diagram?

3 Takis spins the spinner twice.

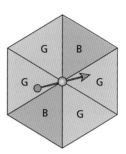

a) Copy and complete the sample space diagram to show all the possible outcomes.
Use it to find the probability that he gets the same colour on both spins.

Second spin

		G	G	G	G	B	B
	G	GG	GG	GG			
	G	GG	GG	GG			
First spin	G	GG					
	G	GG					
	B	BG					
	B	BG					

b) Copy and complete the tree diagram to show the outcomes from each spin.
Use it to find the probability that he gets blue on both spins.

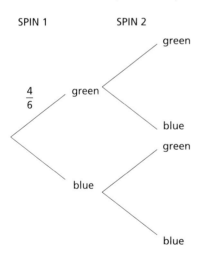

SPIN 1   SPIN 2

green

$\frac{4}{6}$ green

blue

green

blue

blue

4 Mona has two bags of sweets. Each bag contains red and orange sweets.
She picks a sweet from each bag at random.

The probability of picking a red sweet from Bag A is 0.7

The probability of picking a red sweet from Bag B is 0.6

a) Copy and complete the tree diagram to show the probabilities.

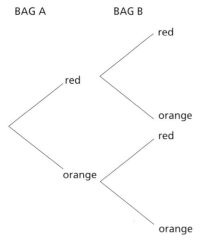

BAG A   BAG B

red

red

orange

red

orange

orange

b) Find the probability that Mona picks two red sweets.

c) Find the probability that she picks two orange sweets.

5 The probability that Hannah has chips for dinner is 0.2

The probability that Hannah has ice cream for dinner is 0.6

Assume that having ice cream for dinner is independent of having chips for dinner.

a) Copy and complete the tree diagram to show the possible outcomes. Write a probability on each branch.

b) Find the probability that she has chips and ice cream.

c) Find the probability that she has chips but does not have ice cream.

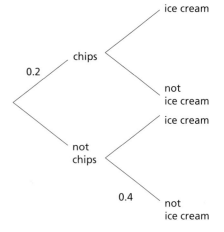

0.2

chips

ice cream

not ice cream

not chips

ice cream

0.4

not ice cream

6 Zak and Camilla each design a spinner.

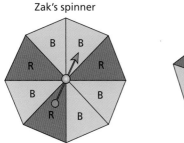

Zak's spinner

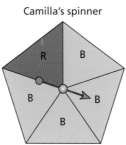

Camilla's spinner

They each spin their spinners.

a) Draw a tree diagram to show the outcomes from each of their spins. Write a probability on each branch.

b) Find the probability that they both spin red.

c) Find the probability that they spin different colours.

7 Paddy throws an ordinary dice twice.

a) Copy and complete the tree diagram to show possible outcomes for each throw.

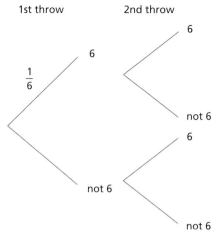

b) Find the probability that he gets a six on both throws.

c) Find the probability that he gets at least one six.

8 Serena chooses one of the cards below at random. She also spins the spinner.

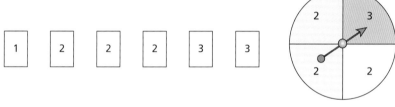

a) Copy and complete this tree diagram.

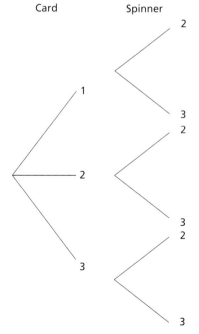

b) Find the probability that both the card and spinner show 2.

c) Find the probability that the card and the spinner show the same number.

d) Find the probability that the sum of the numbers on the card and spinner is 5.

9 Jack wears either a white or a blue shirt each day to work.
He is equally likely to wear each colour of shirt.

He also wears either blue or black trousers. He is twice as likely
to wear blue trousers as black trousers.

a) Draw a tree diagram to show the possible outcomes.

b) Jack says that the probability that he wears a blue shirt and blue trousers is $\frac{1}{2} + \frac{2}{3} = \frac{7}{6}$

  i) How can you tell from his answer that his calculation **must** be wrong?

  ii) Explain what mistake Jack has made in calculating the probability.

c) Find the probability that Jack's shirt and trousers are different colours.

Thinking and working mathematically activity

Hira and Ingrid each have a spinner. Each of their spinners contain some red sections and some blue sections.

They both spin their spinners.

* The probability that both spinners land on a red section is $\frac{4}{25}$.
Draw a possible spinner for each person.

* What if the probability of both spinners landing on red is $\frac{1}{2}$?
Draw possible spinners in this case.

Consolidation exercise

1 Max takes a counter at random from this bag.

He puts the counter back in the bag.
He then takes out a second counter at random.

a) Decide if the colour of the first counter is independent
of the colour of the second counter. Give a reason.

b) Find the probability that on both picks he takes a counter
with the number 4.

2 Lottie buys a drink and a biscuit from a café.
The probability she buys a hot drink is 0.55
The probability she buys a chocolate biscuit is 0.8

a) Calculate the probability that she buys a hot
drink and a chocolate biscuit.

b) Copy and complete the tree diagram by
writing in all the probabilities.

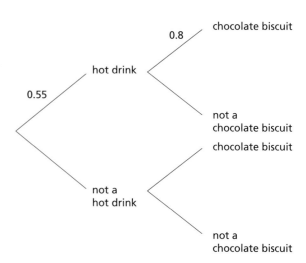

3 Sam and Tia play a game. They spin each spinner and add together the scores.

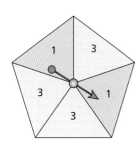
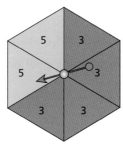

a) Draw a tree diagram to show the possible numbers on the spinners.

b) Sam wins if the total score is 6. Otherwise Tia wins.

Sam says that the game is fair as they both have the same chance of winning.

Is he correct? Show how you worked out your answer.

4 Yannis has a biased coin. The probability that it lands on a head is 0.6

He spins the coin twice.

a) Draw a tree diagram to show the outcomes. Include the probabilities on each branch.

b) Calculate the probability of spinning two tails.

c) Calculate the probability of spinning exactly one head.

5 Kira takes a card at random from each set of cards.

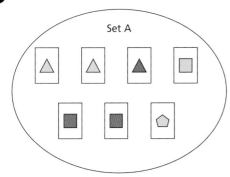
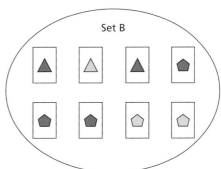

a) Copy and complete the gaps to make the calculation match the statement:

The probability of picking a red .... from set A and a .... triangle from set B

is $\frac{2}{7} \times \frac{1}{8} = \frac{1}{28}$

b) Draw a tree diagram to show the possible colours of the shapes picked from each set.
Use the tree diagram to find the probability that she picks shapes with matching colours.

c) By drawing a different tree diagram, find the probability that the card picked from
Set A is a shape with fewer sides than the card picked from Set B.

6 A bag contains four red balls, three green balls and one blue ball.

Chloe takes a ball from the bag at random.

She puts the ball back in the bag and then takes out a second ball.

a) Chloe writes a correct statement but she spills ink over one word.

The probability that I get a ⬛ ball on each go is $\frac{9}{64}$

Find the word that the ink spill has covered up.

b) Find a way to complete this statement:

The probability that Chloe takes .... is $\frac{24}{64}$

End of chapter reflection

You should know that...	You should be able to...	Such as...
Events are independent if the probability of one event happening does not depend on whether or not the other happens.	Identify if two events are independent.	Explain why these may not be independent: A: having spaghetti for lunch today B: having spaghetti for lunch tomorrow
If events A and B are independent, $P(A \text{ and } B) = P(A) \times P(B)$	Find the probability of two events both happening.	A dice is thrown and a coin is spun. Find the probability of getting a number greater than 4 and a tail.
Probabilities on the branches of a tree diagram add to 1.	Use a tree diagram to find the probability of a combined event.	A bag has four red balls and a green ball. A ball is taken out at random and then replaced. A second ball is then taken out. Draw a tree diagram to show the possible outcomes. Use it to find the probability of getting balls of the same colour.

24

3D shapes

You will learn how to:

- Use knowledge of area and volume to derive the formula for the volume of prisms and cylinders. Use the formula to calculate the volume of prisms and cylinders.
- Use knowledge of area, and properties of cubes, cuboids, triangular prisms, pyramids and cylinders to calculate their surface area.
- Identify reflective symmetry in 3D shapes.

Starting point

Do you remember...

- how to find the area of 2D shapes?

  For example, find the area of the circle and trapezium.

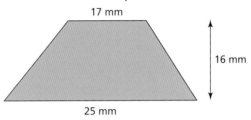

- how to find the volume of cuboids and triangular prisms?

  For example, find the volume of these 3D objects.

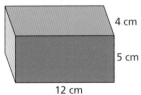

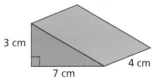

- how find the surface area of objects using nets?

  For example, find the surface area of the square-based pyramid with this net.

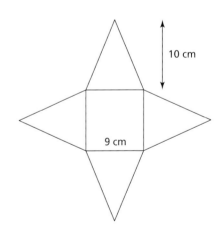

- how to identify all the symmetries of 2D shapes?

  For example, describe all the symmetries of this shape.

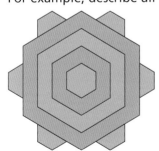

This will also be helpful when...

- you find the volume and surface area of other prisms and more complex 3D shapes.

24.0 Getting started

You will need ten cubes and isometric paper and a partner for this activity.

Arrange five cubes to make a 3D object.

Ask your partner to create the mirror image of your object.

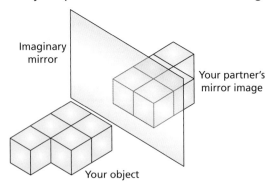

Swap over roles so that your partner makes an object first then you create the mirror image.

Draw your shapes and images on isometric paper.

24.1 Volume of prisms

Key terms

The ends of a **prism** are the same shape. This same shape is the **cross section** that runs through the whole prism.

The volume of a prism can be found using the formula:

volume of prism = area of cross section × length

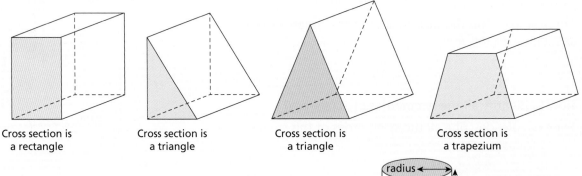

Cross section is a rectangle   Cross section is a triangle   Cross section is a triangle   Cross section is a trapezium

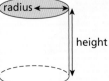

A **cylinder** is an object with a circular cross section.

The formula for the volume of a cylinder = πr^2 × height = $\pi r^2 h$.

Look at this cylinder. It has been cut into 16 slices.

Imagine all these slices being rearranged to approximately make a cuboid:

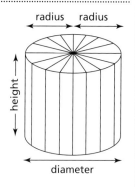

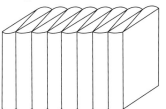

What is the length, width and height of the cuboid?

Can you see why the formula for the volume of a cylinder is $\pi r^2 h$?

Worked example 1

Find the volume of each solid.

a)

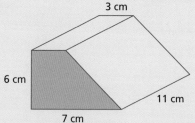

3 cm

6 cm

11 cm

7 cm

b)

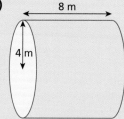

8 m

4 m

a) Area of trapezium	First work out the area of the cross section of the prism. Use the formula for the area of a trapezium:	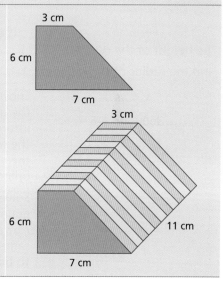
$= \frac{1}{2} \times (3 + 7) \times 6$		3 cm
$= \frac{1}{2} \times 10 \times 6$	Area $= \frac{1}{2}(a + b)h$	6 cm
$= 30 \text{ cm}^2$		7 cm
Volume $= 30 \times 11$	Substitute the area and length into the formula for the volume of a prism:	3 cm
$= 330 \text{ cm}^3$		6 cm
	volume = area of cross section × length	11 cm
		7 cm

b) $V = \pi r^2 h$

$V = \pi \times 4^2 \times 8$

$= 402.1 \text{ m}^3$ (1 d.p.)

Write down the formula for the volume of a cylinder.

Substitute the radius and the height measurements into the formula.

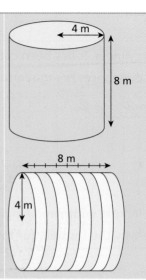

Worked example 2

A cylinder has a volume of 360 cm³.

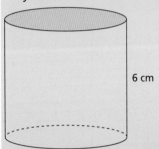

6 cm

The height of the cylinder is 6 cm.

Find the radius of the cylinder.

$V = \pi r^2 h$

$360 = \pi r^2 \times 6$

$\div 6 \qquad \div 6$

$60 = \pi r^2$

$\div \pi \qquad \div \pi$

$19.098... = r^2$

$r = \sqrt{19.098....}$

$= 4.37 \text{ cm}$

The radius is 4.37 cm (3 s.f.)

Substitute the given information into the formula for the volume of a cylinder.

Solve the equation to find r by applying the same operation to both sides.

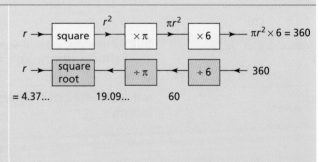

1 Match each shape with its volume.

| Volume = 480 cm³ | Volume = 660 cm³ | Volume = 252 cm³ | Volume = 280 cm³ |

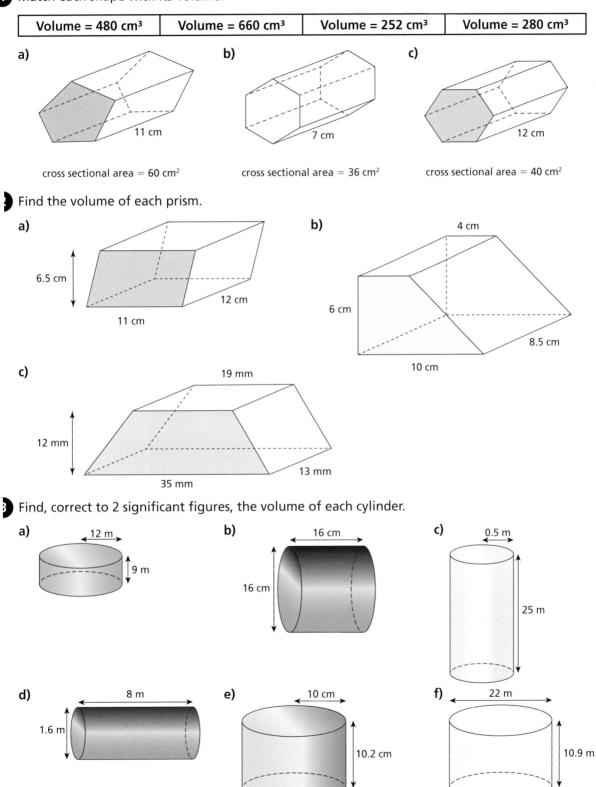

a)
11 cm

cross sectional area = 60 cm²

b)
7 cm

cross sectional area = 36 cm²

c)
12 cm

cross sectional area = 40 cm²

2 Find the volume of each prism.

a)
6.5 cm
12 cm
11 cm

b)
4 cm
6 cm
8.5 cm
10 cm

c)
19 mm
12 mm
13 mm
35 mm

3 Find, correct to 2 significant figures, the volume of each cylinder.

a)
12 m
9 m

b)
16 cm
16 cm

c)
0.5 m
25 m

d)
8 m
1.6 m

e)
10 cm
10.2 cm

f)
22 m
10.9 m

4 Show that the volumes of these two solids are the same.

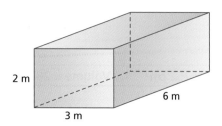

2 m
6 m
3 m

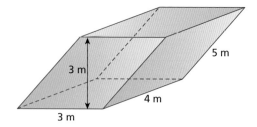

3 m
5 m
4 m
3 m

5 Use the given information to find the missing lengths from the diagrams below.
Choose your answers from the numbers in the box.

| 10 cm | 9 cm | 9 cm (1 s.f.) | 5 cm | 9.5 cm |

a)

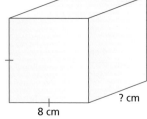

8 cm
? cm

Volume = 608 cm³

b)

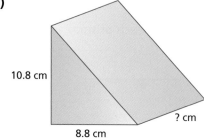

10.8 cm
? cm
8.8 cm

Volume = 475.2 cm³

c)

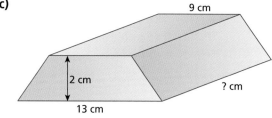

9 cm
2 cm
? cm
13 cm

Volume = 110 cm³

d)

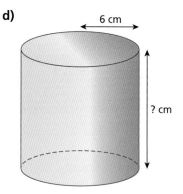

6 cm
? cm

Volume = 1018 cm³

6 This prism has a right-angled triangle as its cross section.

This prism has a volume of 260 cm³.

Zahara is finding the missing length.

She writes:

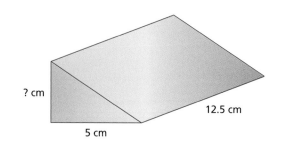

? cm
12.5 cm
5 cm

Volume = area of triangle × height
Area of triangle = base × height ÷ 2
Area of triangle = 260 ÷ 12.5 = 20.8 cm²
Height = 20.8 ÷ 5 ÷ 2 = 2.08 cm

Do you agree with Zahara? Explain your answer.

7 A cylinder has a volume of 64 cm³ and a height of 8 cm. Find the diameter of the cylinder. Give your answer to 2 significant figures.

8 This shape has a semicircular cross section. Calculate the volume. Give your answer to 3 significant figures.

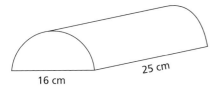

16 cm   25 cm

Thinking and working mathematically activity

The prism below has a volume of 72 cm³.

All the dimensions are integers.

• List the possible pairs of missing values.

• If the height (3 cm) was changed to a different integer value, what possible values could you now have if the volume stays at 72 cm³?

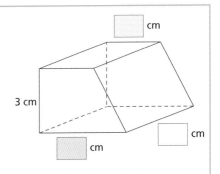

3 cm

cm

cm

cm

24.2 Surface area

Key terms

The **surface area** of a prism can be found by adding together the area of each face.

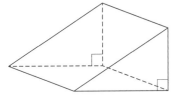

The **surface area** of this triangular prism can be found by adding the area of the two triangular faces and the area of the three faces that are rectangles.

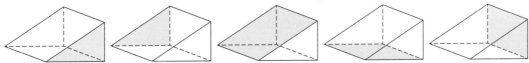

The surface area of a cylinder can be found by adding the area of the top and bottom circle and the area of the curved surface.

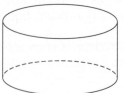

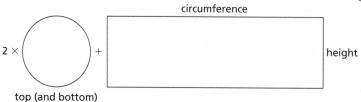

circumference

$2 \times$ + height

top (and bottom)

Worked example 3

Calculate the surface area of each shape.

a)
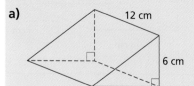
12 cm
6 cm
8 cm

b)
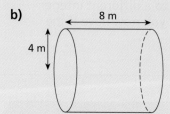
8 m
4 m

		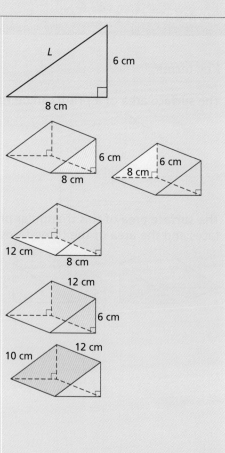

a) $L^2 = 8^2 + 6^2$

$= 100$

$L = \sqrt{100} = 10$ cm

To find the area of the sloping rectangle, you need to know the length of the hypotenuse.

This can be found using Pythagoras' theorem.

Area of triangle

$= \frac{1}{2} \times base \times length$

$= \frac{1}{2} \times 6 \times 8$

$= 24$ cm²

Find the area of the triangular faces.
Substitute the base and height measurements into the area formula.

Area of base

$= 8 \times 12 = 96$ cm²

The base is a rectangle so find the area (length × width).

Area of side face

$= 6 \times 12 = 72$ cm²

The side face is a rectangle so find the area (length × width).

The sloping face is a rectangle so find the area (length × width) using the value calculated for L.

Area of sloping face

$= 10 \times 12 = 120$ cm²

Find total area of the two triangles and the three rectangles.

Total surface area

$= 2 \times 24 + 96 + 72 + 120$

$= 336$ cm²

b) Area of one circle = πr^2
            = $\pi \times 4^2$
            = 16π
            = 50.265 m² (3 d.p.)

Find the area of one the circular ends.

Write down the answer to at least 2 decimal places.

Circumference of circle = πd
        = $\pi \times 2 \times 4$
        = $8 \times \pi$
        = 25.133 m (3 d.p.)

The curved surface of a cylinder is a rectangle with the same length as the circumference of the circle. Work this out first.

Area of rectangle
        = 23.133 × 8
        = 201.062 m² (3 d.p.)

Now find the area of the curved surface (length × height)

Total area
    = (2 × 50.265) + 201.062
    = 301.59…. m²
    = 301.6 m² (1 d.p.)

Add together the area of two ends and the curved surface.

Write down the final answer to 1 decimal place.

(diagram labels: 4 m, Circumference, 8 m, 25.133 m, 8 m)

Did you know?

Every year we dump over 2 billion tonnes of waste. If all this waste was put on trucks and put in a line, the line would go around the world 24 times.

Calculating the correct surface area is very important in packaging. These days more effort is being put into minimising packaging so the waste is kept to a minimum.

Exercise 2

1 **a)** Here are four prisms. Calculate their surface areas.

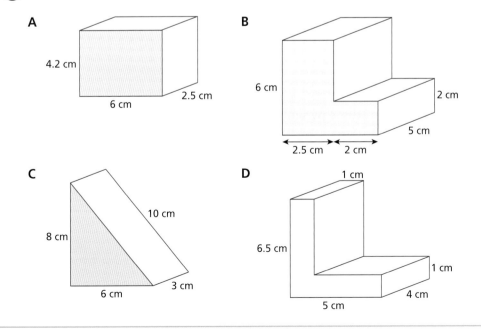

A   4.2 cm, 6 cm, 2.5 cm

B   6 cm, 2 cm, 5 cm, 2.5 cm, 2 cm

C   10 cm, 8 cm, 6 cm, 3 cm

D   1 cm, 6.5 cm, 1 cm, 5 cm, 4 cm

b) Copy the Venn diagram. Write the the labels, A, B, C and D, of the four prisms in the correct position on the diagram.

Surface area of prisms

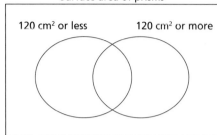

120 cm² or less   120 cm² or more

2 The diagram shows a prism. The cross section is a right-angled triangle.

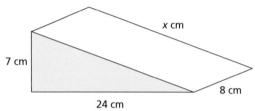

x cm

7 cm

8 cm

24 cm

a) Use Pythagoras' theorem to calculate the value of *x*.

b) Calculate the total surface area of the prism.

3 Calculate the surface area of each of these pyramids. In each pyramid, the top vertex is vertically above the centre of the base. All units are cm.

a)

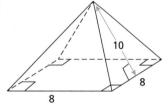

10

8

8

b)

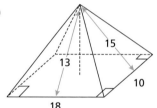

15

13

10

18

Did you know?

A pyramid which has the top vertex vertically above the centre of the base is known as a right pyramid.

Think about

How would you find the surface area of these pyramids? The units are cm.

a)

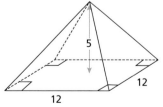

5

12

12

b)

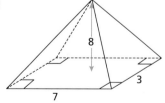

8

7

3

4 Here are four cylinders. Match the cylinders that have equal surface areas.

a)

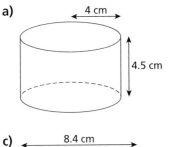
4 cm
4.5 cm

b)

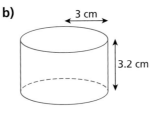
3 cm
3.2 cm

c)

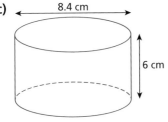
8.4 cm
6 cm

d)

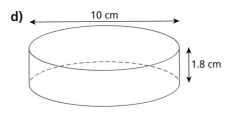
10 cm
1.8 cm

5 Helena's pencil pot is in the shape of a cylinder.

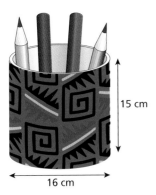
15 cm
16 cm

She decorates the surface area of the pot including the base. What is the pot's total (outside) surface area?

> **Tip**
>
> There is no face on the top.

6 Ben thinks the surface area of this prism is equal to the sum of the surface area of the two cuboids. Is he correct? Show your working clearly.

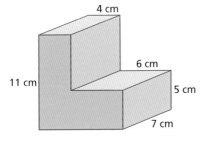
4 cm
6 cm
11 cm
5 cm
7 cm

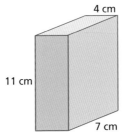
4 cm
11 cm
7 cm

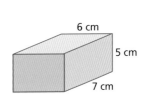
6 cm
5 cm
7 cm

7 These shapes have an equal surface area.

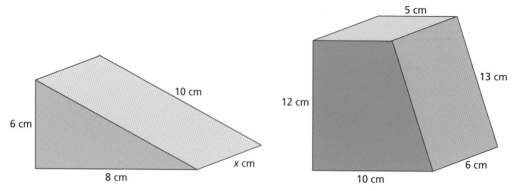

Calculate the value of x.

8 This shape is a prism with a trapezium cross section.

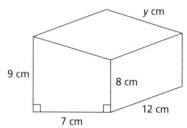

a) Use Pythagoras' theorem to calculate the value of y. Give your answer to 2 decimal places.

b) Calculate the total surface area of the prism. Give your answer to the nearest cm^2.

▼ Thinking and working mathematically activity

Two different cylinders have equal volumes.

Ollie says, 'Their surface areas are equal too.'

Investigate if this is always true, sometimes true or never true.

24.3 Reflective symmetry in 3D shapes

Key terms

....

A **plane of symmetry** divides a 3D shape in two congruent parts which are mirror images of each other.

Worked example 4

a) A prism has a cross section that is an equilateral triangle.

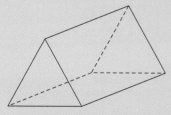

How many planes of symmetry does this prism have?

---

b) Complete the object so that the shaded plane is a plane of symmetry.

a) The prism has a total of four planes of symmetry.

The equilateral triangle has three lines of symmetry.

The prism has a plane of symmetry corresponding to each of these lines of symmetry.

The fourth plane of symmetry cuts the prism in half parallel to the end faces.

b)

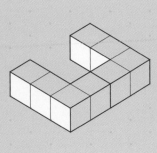

The original diagram shows four cubes. So four more cubes must be added on the opposite side of the plane of symmetry to make a mirror image.

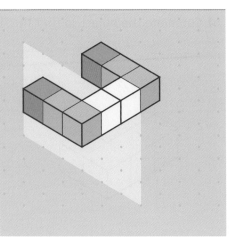

Exercise 3

1 Write down the number of planes of symmetry for each of these 3D shapes.

a) 2 cm by 3 cm by 1 cm cuboid

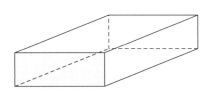

b) prism with a regular pentagon as cross section

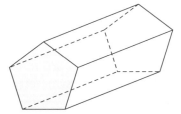

c) prism with semi-circular cross section

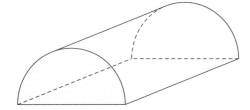

d) prism with trapezium cross section

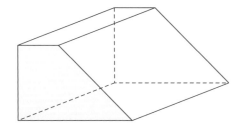

2 Sketch diagrams showing the position of all the planes of symmetry of:

a) a prism with an isosceles triangle as cross section

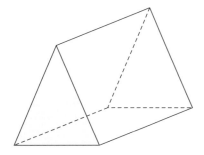

b) a cuboid with a square base

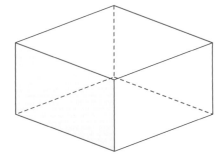

c) a regular octahedron

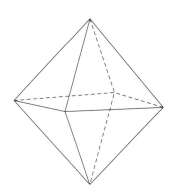

d) a T-shaped prism

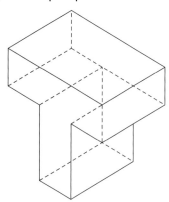

3 Sketch a pyramid with five planes of symmetry.

4 **a)** The cross section of each of these prisms is a regular polygon.

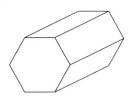  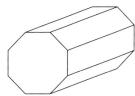

Write down the number of planes of symmetry for each prism.

b) How many planes of symmetry will there be in a prism with a regular 100-sided polygon as its cross section?

c) Find a relationship between the number of sides of the regular polygon and the number of planes of symmetry in the prism.

5 Copy each of these diagrams onto isometric paper and complete each object so that the plane is a plane of symmetry.

a)

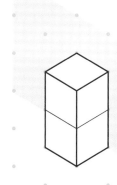

b)

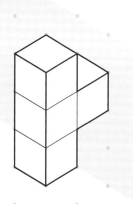

c)

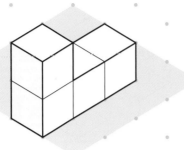

d)

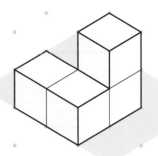

6 A teacher asks Vasili to draw the planes of symmetry on this rectangular based pyramid. The vertex is vertically above the centre of the base.

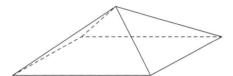

Here is Vasili's answer.

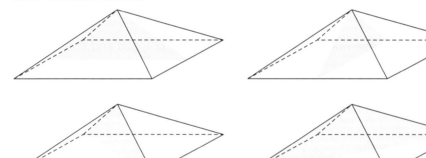

Has Vasili marked the planes of symmetry on the pyramid correctly?

Explain your answer.

7 Technology question Use drawing software to design a building which possesses a vertical plane of symmetry.

Thinking and working mathematically activity

Investigate the number of planes of symmetry in objects made from six cubes. Try to find examples of objects made from six cubes that have different numbers of planes of symmetry. Draw your objects on isometric paper. Think of ways that you could extend your investigation.

---

1 Calculate the volume and surface area of this cylinder.

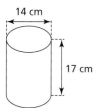

14 cm

17 cm

2 Which two of these shapes have exactly two planes of symmetry?

a)

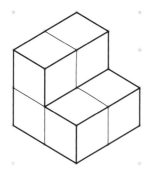

b)

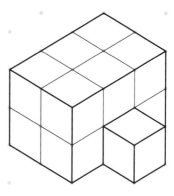

c)

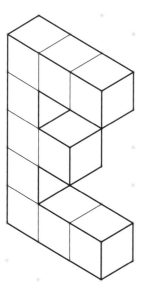

d)

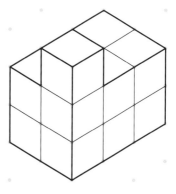

Vocabulary question Match the key term to its description or formula.

Area of a circle	The total area of all the faces of a 3D shape
Circumference of a circle	$\pi r^2 h$
Prism	Area of cross section $\times$ length
Surface area	$2\pi r^2 + 2\pi r \times h$
Volume of a prism	The length around a circle $= \pi d$ or $2\pi r$
Curved surface area of a cylinder	πr^2
Surface area of a cylinder	A 3-dimensional shape with a fixed cross section
Volume of a cylinder	Circumference $\times$ height of the cylinder

The depth of this water in this tank is 80 cm. Calculate the volume of the water in the tank.

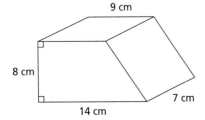

3 m

2.3 m

Decide if these statements are true or false. Give a reason for your answer.

Statement 1: All prisms have at least one plane of symmetry.

Statement 2: No prism has exactly 2 planes of symmetry.

Statement 3: The maximum number of planes of symmetry for objects made from 4 cubes is 4.

6 A tunnel, in the shape of a semicircle, is cut through a hillside.

The diameter of the semicircle is 15 m.

The length of the tunnel is 240 m.

Find the total volume of material removed from the hillside.

Calculate **a)** the volume and **b)** the total surface area of this prism.

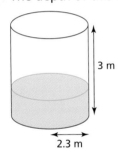

9 cm

8 cm

14 cm

7 cm

8 A container is made as a prism.

The top is made of a rectangle 5 cm by 9 cm with a semicircle at each end.

The depth is 3 cm. Calculate the volume of the container.
Give your answer to 1 decimal place.

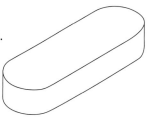

End of chapter reflection

You should know that...	You should be able to...	Such as...
Volume of a prism = area of cross section × length. Volume of a cylinder = $\pi r^2 h$	Calculate volumes in prisms and cylinders.	Calculate the volume of this cylinder. 6.5 cm 12 cm
	Calculate a missing length in prisms and cylinders given the volume.	The volume of the prism = 140 cm³. Calculate the missing length. 3 cm 7 cm ? cm
The surface area of a 3D shape is the total of the areas of each of its faces.	Calculate the surface area of prisms and cylinders. Calculate a missing length in prisms and cylinders given the surface area.	Calculate the surface area of the triangular prism. 29 cm 21 cm 35 cm 20 cm
A plane of symmetry divides a 3D shape into two halves that are mirror images of each other and congruent.	Identify reflective symmetry in 3D shapes.	Write down the number of planes of symmetry in this prism.

Simultaneous equations

You will learn how to:
- Understand that the solution of simultaneous linear equations:
  - is the pair of values that satisfy both equations
  - can be found algebraically (eliminating one variable)
  - can be found graphically (point of intersection).

Starting point

Do you remember...

- how to solve a linear equation, including when the solution is a negative number or decimal?

  For example, solve $4x - 8 = 17$
- how to calculate with negative numbers?

  For example, work out $2 - (-7)$
- how to substitute into a formula to find a missing value?

  For example, if $3x + 2y = 18$ and $x = 2$, what is y?
- how to rearrange a simple formula?

  For example, make x the subject of $x + y = 6$
- how to plot linear graphs?

  For example, plot the graph of $y = 3x - 2$

This will also be helpful when...

- you learn more about solving equations that represent real-life situations.

25.0 Getting started

Here is a game for two players:

Player 1 takes two dice and rolls them so that Player 2 cannot see the numbers on the dice.

Player 1 then tells Player 2 the total of the two numbers.

Player 2 has a guess at what the two numbers are.

If Player 2 is correct on the first guess, they score 3 points.

If Player 2 is not correct, Player 1 must tell Player 2 the difference between the two numbers on the dice.

Player 2 has a second chance to guess the two numbers.

If Player 2 is correct, they score 1 point.

The players then swap roles until they have both had five turns.

Try playing this game with a partner.

Now, imagine that you are Player 2.

Player 1 tells you that the dice have a total of seven.

What could the numbers on the dice be?

Player 1 then tells you that the difference between the scores is 3. What must the scores be?

You have just solved a pair of simultaneous equations!

Key terms

Simultaneous equations are a set of two (or more) equations that contain the same unknowns.

For example,

$x + 5y = 15$

$x + 3y = 11$

When you **solve** simultaneous equations, you find the values of the unknowns that satisfy both (or all) of the equations at the same time.

One way that you can solve simultaneous equations is by **eliminating** or removing one of the unknowns to give an equation with just one unknown, which you can then solve. You can then **substitute** this solution into one of the original equations to find the value of the second unknown.

If a pair of simultaneous equations have the same number of one of the unknowns in both equations, then they are said to have **like coefficients**.

For example, in the equations $\quad x + 2y = 12 \quad$ the y terms have like coefficients.

$\qquad\qquad\qquad\qquad\qquad 3x + 2y = 16$

Worked example 1

Solve the simultaneous equations:

$x + 5y = 15$

$x + 3y = 11$

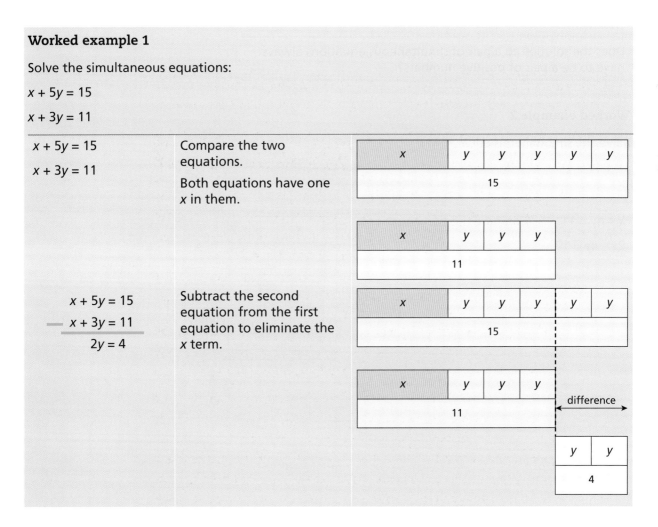

| $x + 5y = 15$ | Compare the two equations. |
| $x + 3y = 11$ | Both equations have one x in them. |

$\quad\ x + 5y = 15$	Subtract the second
$-\ \underline{x + 3y = 11}$	equation from the first
$\qquad\ \ 2y = 4$	equation to eliminate the x term.

$y = 2$

$x + 5y = 15$

$x + 5 \times 2 = 15$

$x + 10 = 15$

$x = 5$

Find the value of y.

Substitute $y = 2$ into one of the original equations to solve for x.

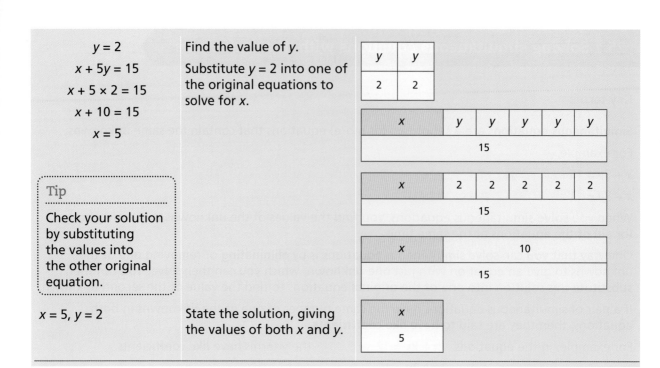

Tip

Check your solution by substituting the values into the other original equation.

$x = 5, y = 2$

State the solution, giving the values of both x and y.

Discuss

Does the solution to a pair of simultaneous equations always have to be a pair of positive numbers?

Worked example 2

Solve the simultaneous equations:

$x + 4y = 11$

$2x - 4y = 10$

$x + 4y = 11$ $2x - 4y = 10$	Compare the two equations.

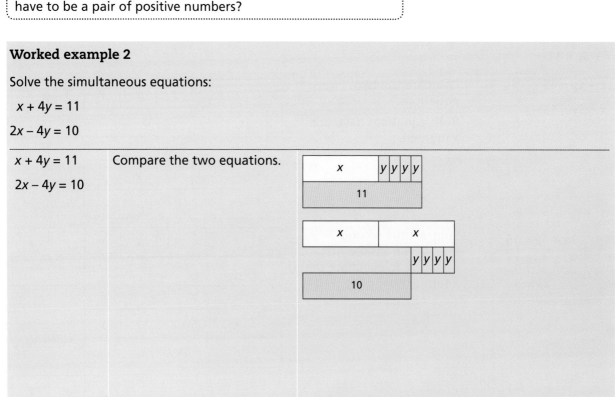

2 Match these simultaneous equations with their solutions.

| $x + y = 5$ |
| $3x + 2y = 12$ |

| $x + y = 1$ |
| $4x + 3y = 1$ |

| $2x = 1$ |
| $4x + 3y = 11$ |

| $x = \dfrac{1}{2}, y = 3$ |

| $x = 2$ |
| $y = 3$ |

| $x = -2$ |
| $y = 3$ |

3 Solve these simultaneous equations using substitution.

a) $y = 3x - 4$
  $y = 2x + 5$

b) $x = 2y + 1$
  $3x + y = 31$

c) $2x + y = 7$
  $x + 3y = 1$

4 Two numbers, a and b, add up to 20 and the difference between them is 4. a is larger than b.

a) Form two simultaneous equations to represent this problem.

b) Solve your equations to find the value of a and b.

5 Jacob and Jon bought some clothes from a sports shop.

Jacob bought two T-shirts and a pair of shorts. He spent $50.

Jon bought one T-shirt and one pair of shorts and spent $32.50

All the T-shirts were the same price.

a) Form two equations to show what they bought and spent.

b) Solve your equations to find the cost of a T-shirt and a pair of shorts.

Thinking and working mathematically activity

Think about this problem:

Erika bought 20 sweets for $2.42.

She bought lemon sweets and strawberry sweets only.

She bought two more strawberry sweets than lemon sweets.

How may of each kind of sweet did Erika buy?

Can you create a pair of simultaneous equations to solve this problem?

> Tip
>
> Use x for the number of lemon sweets and y for the number of strawberry sweets.
>
> Solve your equations and check that your solution works for the original problem.
>
> Can you make up another problem like this to challenge a partner?

Key terms

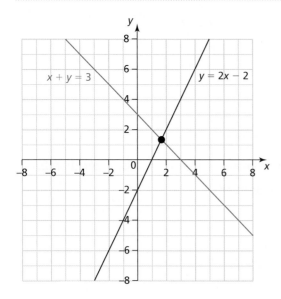

You can also solve simultaneous equations using graphs. If you draw graphs of both equations on a coordinate grid, the **point of intersection,** where the two lines cross, is where the values of x and y satisfy both equations, and so gives the solution to the equations.

For example, this diagram shows that the solution to the simultaneous equations :

$x + y = 3$

$y = 2x - 2$

is the point where the lines cross, which gives the solution $x = \dfrac{5}{3}$ and $y = \dfrac{4}{3}$

Worked example 5

Solve the equations $y = 3 - x$ and $y = 4x - 2$ graphically.

$y = 3 - x$ $y = -x + 3$ Gradient is -1 y-intercept is 3. $y = 4x - 2$ Gradient is 4 y-intercept is 2. The point of intersection is $(1, 2)$ so $x = 1$ and $y = 2$.	Rearrange $y = 3 - x$ into the form $y = mx + c$. Plot the graphs of $y = 3 - x$ and $y = 4x - 2$ either by drawing a table of values or by finding the gradient and intercept of each line.	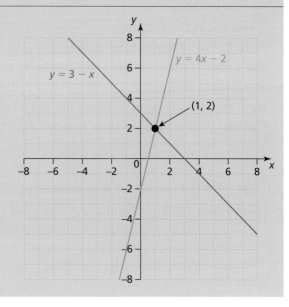

Tip

The point where the lines cross will always be the solution to the simultaneous equations, as that is the only pair of values that is on both lines, so satisfy both equations.

▶ Use the graphs to write down the solutions to each pair of simultaneous equations.

a) $y = x$ and $x = 5$

b) $y = x$ and $2x + y = 6$

c) $2x + y = 6$ and $x = 5$

d) $2y + 3 = x$ and $x = 5$

e) $2y + 3 = x$ and $2x + y = 6$

f) $2y + 3 = x$ and $y = x$

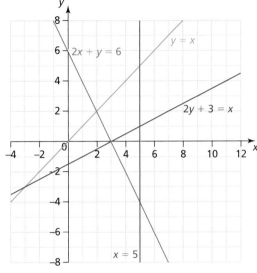

▶ **Technology question** Use graphing software to find the solution to these simultaneous equations.

a) $3x + 2y = 8$ and $x = 2y$

b) $3x + y = 3$ and $y = x + 1$

c) $2x = 3y + 2$ and $2x = 4y + 3$

d) $4x + 5y = 17$ and $2y = 6x + 3$

> **Tip**
>
> Some of the graphs may not intersect at whole number coordinates.

3 a) Copy and complete the table of values for $x + 2y = 11$ and $2x + y = 13$.

$x + 2y = 11$

x	0		7
y		0	

$2x + y = 13$

x	0		4
y		0	

b) Plot the graphs of both lines on the same set of axes.

c) Find the solutions to the simultaneous equations:

$x + 2y = 11$

$2x + y = 13$

4 **a)** Copy and complete the table of values for $y - x = 4$ and $2x + y = 7$.

$y - x = 4$

x	−1	0	2
y			

$2x + y = 7$

x	−1	0	2
y			

b) Plot the graphs of both lines on the same set of axes.

c) Find the solutions to the simultaneous equations:

$y - x = 4$

$2x + y = 7$

Think about

A pair of simultaneous equations have solutions $x = 1$, $y = -1$.
What possible pairs of equations have these solutions?

---

5 Savanah says the simultaneous equations $2y + x = 3$ and $y = 5 - 0.5x$ have no solution.
Draw the graphs for these equations and use your diagram to help to explain your answer.

6 Ruby buys three T-shirts and two pairs of sandals for $31.
Donna buys four T-shirts and three pairs of sandals for $44.

a) Write a pair of simultaneous equations to show what they bought.

b) Solve the simultaneous equations by plotting the graph of both equations on one set of axes.

Thinking and working mathematically activity

Steve is trying to solve these sets of simultaneous equations but cannot seem to find any solutions.

a) $y = 2x - 5$ **b)** $x + 2y = 4$

    $6x = 3y + 15$ $y = \frac{1}{2}x + 5$

Use tables of values and graphs to explain to Steve why he cannot solve these equations.
Make up some more similar simultaneous equations to test a partner.

Consolidation exercise

1 Solve these pairs of simultaneous equations:

a) $3x + y = 15$ **b)** $x + y = 11$ **c)** $3x + 2y = 16$

    $2x + y = 11$ $4x - y = 19$ $x - y = 2$

Match each pair of simultaneous equations with their solution, completing the empty boxes with a suitable equation or solution where necessary.

A
$$x + 5y = 8$$
$$x + 2y = 5$$

V
$$x = 5$$
$$y = -3$$

B
$$5x - 2y = 31$$
$$7x + 2y = 29$$

W

C

X
$$x = 3$$
$$y = 1$$

D
$$3x + 3y = 18$$
$$9x - 2y = 43$$

Y
$$x = 5$$
$$y = 1$$

E
$$2x + y = -7$$
$$4x - 3y = 1$$

Z
$$x = 1$$
$$y = -3$$

3 Choose the pair of equations from the box which have solutions $x = -2$ and $y = 3$.

$$2y = x + 3$$ $$3x + 2y = 0$$
$$y - x = 4$$ $$y = x + 5$$

4 Solve these equations by using the method of substitution.

a) $y = x + 1$
   $x + y = 13$

b) $x = 4 - 2y$
   $2x + 3y = 9$

c) $y = 5 - 3x$
   $2y = 4x + 15$

5 a) Copy and complete the tables of values for $y = 2x + 1$ and $x + 2y = 12$.

$y = 2x + 1$

x	−1	0	1	3
y				

$x + 2y = 12$

x	−1	0	1	3
y				

b) Plot the graphs of both lines on the same set of axes.

c) Use your graph to solve the simultaneous equations

$y = 2x + 1$

$x + 2y = 12$

6 The difference between the size of the two acute angles of a right-angled triangle is 42°.

a) Form a pair of simultaneous equations to show this information.

b) Solve the simultaneous equations to find the size of these two acute angles in degrees.

7 231 people go on a road trip.

Some of the people travel in vans and others in buses

A van can carry 7 people and a bus can carry 25 people.

There are 15 vehicles in total and no spare seats on any vehicle.

a) Write two simultaneous equations to represent this problem.

b) Solve your equations to find how many of each type of vehicle are used.

8 Which of these pairs of simultaneous equations is the odd one out? Explain your answer.

$3x + 2y = 11$ $x - y = -7$ $-3x + y = -17$ $x + 6y = -7$

$x - 3y = 11$ $x + y = 3$ $3x + 4y = 7$ $2x + 5y = 0$

9 Technology question How could you use a spreadsheet to solve a pair of simultaneous equations by trial and improvement?

For example, try solving $3x + y = 21$ and $7x - y = -13$.

10 Cameron says that two lines intersect at (1, 5). One of the lines has the equation $2y + 3x = 13$. What is the equation of the other line? Check your answer using graphing software.

End of chapter reflection

You should know that...	You should be able to...	Such as...
Simultaneous equations are pairs (or sets) of equations with more than one unknown value.	Recognise a pair of simultaneous equations.	Solve: a) $3x + 2y = 16$ b) $x - y = 2$
Simultaneous equations can be solved be eliminating one of the unknown values.	Eliminate x or y by: – subtracting the equations – adding the equations – multiplying one of the equations to get equal coefficients before adding or subtracting – substituting one equation into the other.	Solve the following sets of simultaneous equations: a) $4x + 3y = 24$ 　 $4x + 2y = 20$ b) $5x + 3y = 33$ 　 $2x - 3y = 9$ c) $3x + 2y = 12$ 　 $7x + 6y = 32$ d) $y = 3x - 1$ 　 $5x - y = 3$
If you draw graphs of simultaneous equations, the point of intersection of the graphs gives the solution to the equations.	Draw graphs to represent simultaneous equations and find the point of intersection.	By drawing graphs, find the solution to: $x + y = 2$ $x = 5 + y$

5 Real life data This graph shows the numbers of Americans receiving welfare.

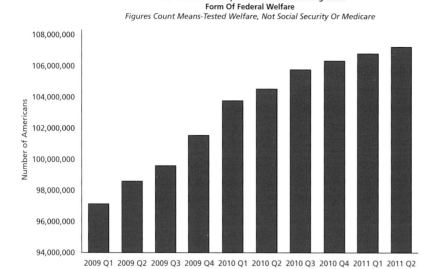

Over 100 Million People In U.S. Now Receiving Some Form Of Federal Welfare
Figures Count Means-Tested Welfare, Not Social Security Or Medicare

Source: The Washington Examiner

a) What is misleading about the graph?

b) What is misleading about the statement 'Welfare payments have more than tripled in two years'?

6 This infographic shows pizza preferences.

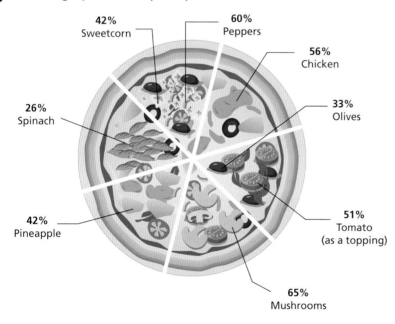

42% Sweetcorn

60% Peppers

56% Chicken

26% Spinach

33% Olives

42% Pineapple

51% Tomato (as a topping)

65% Mushrooms

Others items not depicted include: onions (62%),
beef (36%), chillies (31%) jalapeños (30%), tuna (22%),
anchovies (18%), 2% of people say they only like Margherita pizzas

Write down what is misleading about this infographic.

7 Look at the two coffee jars below:

a) Why is this advertisement misleading?

b) Which package is the best value for money?

8 The graphs show the unit sales for a games console from 2014 to 2018.

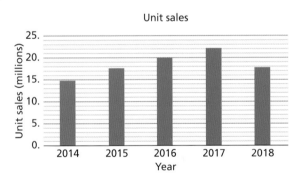

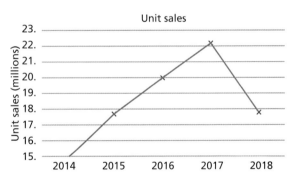

Which of the graphs shows the most accurate representation of the sales?
Give a reason for your answer.

Discuss

Why are some graphs made deliberately misleading?
Can you find some examples in the media?

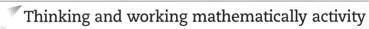

Thinking and working mathematically activity

Here is the population data for the USA from 1960 – 2020.

Use this data to present two graphs one that is fair and one that is misleading. Explain why one of your graphs is fair and the other is misleading.

Year	Population
2020	329 million
2015	320 million
2010	309 million
2005	295 million
2000	282 million
1995	266 million
1990	249 million
1985	237 million
1980	227 million
1975	215 million
1970	205 million
1965	194 million
1960	180 million

26.2 Project

Research project – cars

The table shows some information about the cost of 40 used cars.

Car number	Cost ($)	Age (years)	Distance travelled (km)	Engine size (litres)	CO_2 emissions (g/km)	Colour
1	6300	11	54000	1.3	128	Blue
2	18800	1	6000	1.3	119	Blue
3	29300	2	21000	1.5	151	White
4	10400	6	94000	1.6	94	Silver
5	9600	10	170000	2.2	171	Black
6	17500	3	12000	1.0	117	White
7	37800	1	5000	2.0	120	White
8	3800	14	187000	2.0	215	Grey
9	21100	2	22000	1.5	125	Black
10	6000	10	114000	1.8	155	Red
11	11800	3	68000	1.3	114	Orange
12	11900	4	61000	1.6	98	Black
13	15000	6	72000	2.2	180	Silver
14	3400	15	108000	1.3	139	Black

Car number	Cost ($)	Age (years)	Distance travelled (km)	Engine size (litres)	CO_2 emissions (g/km)	Colour
15	26 100	1	8000	1.0	107	Silver
16	10 000	9	46 000	1.8	169	Red
17	3800	11	118 000	1.3	128	Silver
18	15 600	3	75 000	1.6	115	Grey
19	11 000	4	106 000	1.6	94	Red
20	14 900	5	118 000	1.6	134	Brown
21	32 400	2	13 000	2.0	176	Black
22	20 600	3	45 000	1.5	139	Grey
23	10 600	7	45 000	1.3	128	Blue
24	1800	15	132 000	1.3	137	Blue
25	32 500	1	8000	1.5	151	Red
26	16 900	3	25 000	1.0	114	Red
27	20 300	4	44 000	1.5	133	Blue
28	7300	9	87 000	1.5	117	Silver
29	12 400	5	16 000	1.3	128	Grey
30	8100	7	53 000	1.3	129	Brown
31	29 300	2	21 000	1.5	151	White
32	14 900	5	118 000	1.6	134	Brown
33	3700	11	181 000	2.2	167	Black
34	5600	8	176 000	1.3	105	Brown
35	1100	19	267 000	2.0	203	Green
36	14 700	2	38 000	1.0	110	Grey
37	5200	12	126 000	2.0	200	Black
38	26 900	1	15 000	1.5	135	White
39	4400	10	125 000	1.3	130	Blue
40	27 500	2	20 000	2.0	177	Brown

Look at the data and decide on a question you would like to research.

Some ideas for your investigation would be:

- What variables affect the cost of a used car?
- What is the relationship between engine size and the amount of carbon dioxide (CO_2) emitted?
- Do cars with larger engine sizes travel more miles per year than cars with smaller engines?

You will need to:

- Make a suitable prediction (hypothesis).

  For example, 'Cars with bigger engines have greater carbon dioxide emissions'.
- Decide on which variables from the data set you will need to use.
- Decide if you will use data from all 40 cars or a sample of these.

- Analyse and present your data using charts and tables using a spreadsheet package. For example, draw a scatter diagram of cost versus distance travelled.
- Interpret your results.
- Make conclusions that relate back to your predictions.
- Think about any limitations with the data.

Make suggestions about how you could improve or extend your investigation.

Consolidation exercise

▶ The 3D pie chart shows information about favourite type of movies.

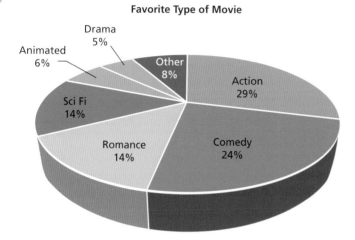

Favorite Type of Movie

What is misleading about this pie chart?

▶ Write down two things that are misleading about this chart.

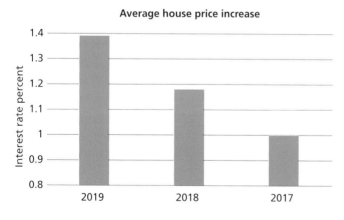

Average house price increase

3 Describe what is misleading about this graph.

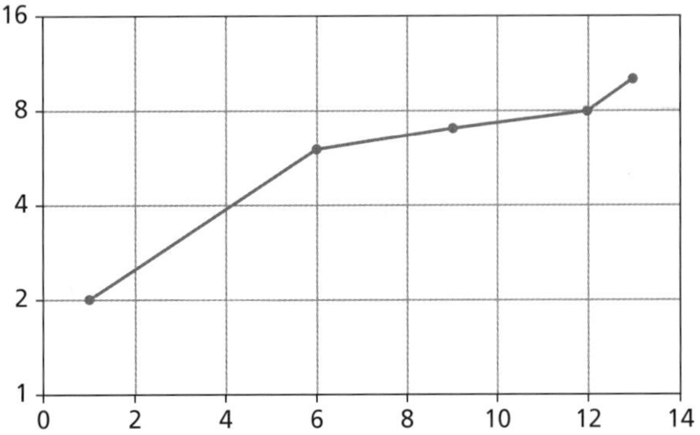

4 The graph shows a company's profits over 3 years.

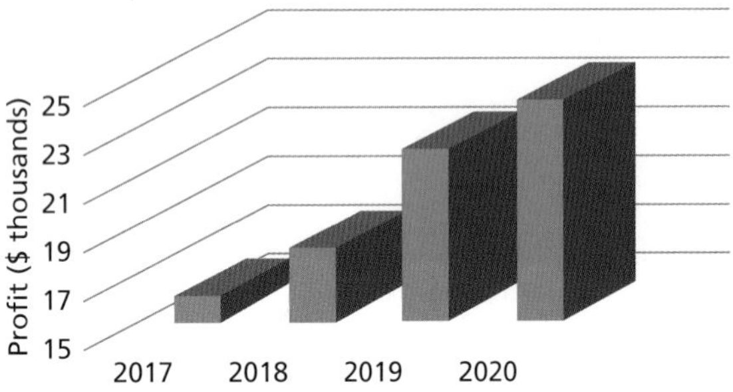

Give two reasons why this graph is misleading.

End of chapter reflection

You should know that...	You should be able to...	Such as...
Graphs can be deliberately drawn to be misleading.	Identify graphs that are misleading and give reasons why.	What is misleading about this chart? 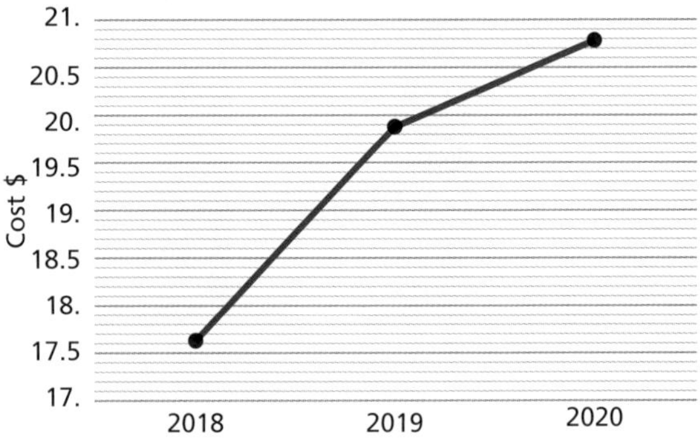